METHUEN'S MONOGRAPHS ON PHYSICAL SUBJECTS

General Editor: B. L. WORSNOP, B.Sc., Ph.D., Head of the Department of Mathematics and Physics, The Polytechnic, and Special Lecturer in Radiology, King's College, London.

F'cap 8vo. Illustrated. Each 2s. 6d. net

THIS series is intended to supply readers of average scientific attainment with a compact statement of the modern position in each subject. The Honours student and the research worker in other branches of physics, those engaged on work in related sciences, and those who are no longer in contact with active scientific work, will find here a series of expositions by authors who are actively engaged in research on the subject of which they write.

A complete list of the above series will be sent on application.

METHUEN & CO. LTD. LONDON

EARTHQUAKES AND MOUNTAINS

GENTLE ANTICLINE WITH CONTINUOUS DISTORTION,
CRICH, DERBYSHIRE

CONTORTION WITH FRACTURE, ASHBOURNE, DERBYSHIRE

EARTHQUAKES AND MOUNTAINS

by
HAROLD JEFFREYS
M.A., D.Sc., F.R.S.

WITH 6 PLATES AND 9 MAPS AND DIAGRAMS

METHUEN & CO. LTD., LONDON
36 Essex Street W.C.2

First published in 1935

PRINTED IN GREAT BRITAIN

PREFACE

This book has been written with the object of providing an account of the modern study of the Physics of the Earth in general language, for the benefit of those interested in natural phenomena and their causes. To a large extent it summarizes my larger book "The Earth" (1929), but attention is given to much work that has been published later. Full references have been given to the latter in case any reader wishes to make a more extended study; it has not been thought necessary to give them for information already discussed more fully in "The Earth," where the earlier references are available. My own work has in the meantime become increasingly concerned with the borderland between geophysics and geology, and geological applications have been emphasized; the astronomical side has been correspondingly reduced.

I acknowledge with thanks the permission of

several authors and publishers to reproduce illustrations from published work; particulars will be found at the respective places.

HAROLD JEFFREYS

St. John's College
 Cambridge

CONTENTS

CHAP.		PAGE
I.	SOLIDS AND LIQUIDS	1
II.	EARTHQUAKES	26
III.	GRAVITY AND THE SHAPE OF THE EARTH	61
IV.	THE STRENGTH OF THE EARTH	84
V.	RADIOACTIVITY AND THE EARTH'S HISTORY	96
VI.	THE BODILY TIDE AND TIDAL FRICTION	134
VII.	THE MECHANICS OF GEOLOGY	144
	INDEX	181

PLATES

Gentle Anticline with Continuous Distortion, Crich, Derbyshire.
Contortion with Fracture, Ashbourne, Derbyshire } *Frontispiece*

                                              FACING PAGE

West Bromwich ($\Delta = 63$ km.) Record of the Hereford Earthquake of 1926 August 14. Showing only P_g and S_g, with a train of oscillations following the latter. A characteristic record at a very short distance   40

(*From Jeffreys 'The Earth,' by courtesy of The University Press, Cambridge.*)

Jena ($\Delta = 481$ km.) Record of the Schwadorf Earthquake of 1927 October 8. Showing the other phases that emerge at intermediate distances   42

(*From Conrad: The Schwadorf Earthquake of 1927 Oct. 8. [Akademische Verlag., m.b.H. Leipzig].*)

Normal Earthquake, 1922 January 31.   48

(*By courtesy of Dr. R. Stoneley and Prof. H. H. Plaskett.*)

Some Records of the Japanese Earthquake of 1931, February 20   50

(*By courtesy of the Royal Society of London and Mr. F. J. Scrase. [Phil. Trans. A699, vol. 231].*)

x   - EARTHQUAKES AND MOUNTAINS

FACING PAGE

CONTINUOUS DISTORTION AND FRACTURE COMBINED,
BROADHAVEN, PEMBROKESHIRE   152

FOLDING IN SHALE DUE TO HILL-CREEP, MATLOCK,
DERBYSHIRE   152

Fig. 4 is based on *F. Kossmat, Geologische Rundschau*, 12, 1921, by permission of the Author.

Figs. 6–9 are from *Geological Magazine*, vol. 68, No. 808, Oct. 1931, by courtesy of the Editors.

EARTHQUAKES AND MOUNTAINS

CHAPTER I

SOLIDS AND LIQUIDS

"... but who attempts to eat an orange without first disposing of the peel, or what manner of a dwelling could be erected unless an adequate foundation be first provided?"
—Ernest Bramah, *Kai Lung's Golden Hours*, 144.

GEOPHYSICS, in its most natural definition, may be held to include everything observable on the earth, and therefore the whole of science. Most of us have had at some stage of our lives the wish to know everything, but in practice we find that some limitation is imposed on us. The reaction to this limitation is that the geophysicist tends to resemble the arm-chair detective. He is provided with a mass of data obtained by numerous independent means, and his task is to co-ordinate them. In some respects, however, he is not so well off as the detective of fiction. As the Ace of Detectives has remarked,* "not only must all prejudices and preconceptions be avoided, but when information is received from outside, the actual undeniable facts must be carefully sifted from the inferences which usually accompany them". It often happens that two different lines of investigations lead to data capable of two inconsistent interpretations;

* R. Austin Freeman, *The Great Portrait Mystery* (*Percival Bland's Proxy*).

but when we consider both sets of data together we are led to a third interpretation different from both, which agrees with both sets of data but disagrees with both the previous interpretations. Unfortunately in geophysics the testing of the third interpretation is usually as difficult as for the first two together. Consequently the geophysicist is in the position of the detective when he has formed a provisional theory and the various clues are falling into their places; but the last chapter, in which the theory leads to conclusions of its own, and these conclusions are all verified, is far from being written.

The facts that show some approach to co-ordination are those of petrology, stratigraphical geology, seismology, gravity determination, earth tides, crustal temperature, and geological time. Meteorology hardly enters into the general scheme; we should expect that the large amount of material obtained in the study of terrestrial magnetism would be relevant to the constitution of the earth, but so far little progress has been made in co-ordinating its results with those of the physics of the earth, as distinct from the atmosphere.

So many of the problems are mechanical that it seems best to begin with some account of the mechanics of solids and fluids, particularly as they involve matters of definition, and different authors sometimes use different definitions. There seems to be no clear-cut distinction between a solid and a fluid. In practice everybody would agree that iron at ordinary temperatures is a solid, and water a liquid; we call treacle

SOLIDS AND LIQUIDS

and tar very viscous liquids; but whether we call glass or pitch a solid or a liquid will depend on what definition we are using. The only procedure to avoid confusion is to state the facts as determined in the laboratory. The fundamental problem is that of the relation between stress and strain. Stress specifies the nature of the internal forces within the material; strain specifies the changes of size and shape arising from those stresses. If we hang a weight on the end of a wire, the wire is stretched and becomes somewhat thinner; the tension within the wire is a stress, the changes of length and thickness are strains. If we compress a ball by a uniform pressure all round it, we reduce its volume; here the pressure is a stress and the reduction of size a strain. If we twist a wire by trying to turn one end one way and the other end the other, there is no change of volume, but there are changes of shape.

Stress is measured by the force per unit area across a surface within the material. In a perfect fluid (which does not exist), or in a real fluid at rest, this force is perpendicular to the surface; but in a solid or in a fluid undergoing changes of shape it may be at any angle to the surface. To specify it completely we therefore need the components of the force in three perpendicular directions. As the surface may be drawn in any direction through a given point, this suggests that the specification of stress may be a matter of great complexity; but actually it can be proved that it is enough to consider only three directions for

4 EARTHQUAKES AND MOUNTAINS

the surface, for if we know the stress-components across three perpendicular surfaces we can find those across any other. Thus nine components are apparently needed to give the stress, but six of them can be shown again to be equal in pairs, leaving only six that are independent. If we take axes at the point parallel to the (perpendicular) axes of x, y, and z, the force per unit area across the plane of x

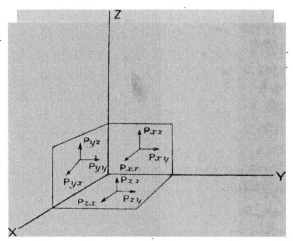

FIG. 1.—Diagram of stress-components across unit areas of three perpendicular planes.

constant tending to pull the matter on the negative side in the direction of x increasing may be denoted by p_{xx}; the force per unit area tending to pull it in the direction of y increasing is p_{xy}, and is equal to the force per unit area of the plane of y constant tending to pull the matter on the negative side in the direction of x increasing, which would naturally be denoted by p_{yx}.

SOLIDS AND LIQUIDS

This is the most general specification of stress possible. But a simplification is possible if we choose the directions of our axes suitably. No matter what the state of stress, three perpendicular directions can always be found such that the stress across a plane containing any two of them is wholly along the third. These stresses are called the principal stresses. Their magnitudes and directions usually vary from point to point of the material. If we take a rectangular block and pull on two opposite faces with equal forces, so as to keep the block at rest, the force per unit area over those faces is a principal stress. We can apply similar pairs of forces over the other faces, and then we have the three principal stresses acting across the faces and at right angles to them. In a perfect fluid the principal stresses are all equal, but in general in real solids and fluids they differ somewhat. If they happen to be equal we say that the stress is *hydrostatic*.

When forces are applied to an element of material it may move in several different ways. Suppose that the element is a rectangular block. The simplest type of motion is that the centre may move; this we call translation. But even if the centre remains where it was the block may rotate without any change in the lengths of the edges. Both types of displacement are possible in a rigid body (which, like a perfect fluid, does not exist). But in addition to these types there may be changes of size or shape; the lengths of the edges may change, as also the angles between them. These changes constitute the *strain*. It may happen that the faces remain rectangles and that all change in the same ratio; then the change is a simple volume expansion or contraction. Sometimes there is no change in volume but there is a change of shape; then we call the strain a pure distortion. In general both volume and shape change. As for stress, the specification can be simplified by a suitable choice of axes. If in any small region we consider rectangular elements with their edges in different directions, they will be distorted in different ways as they move; but we can find three directions such that if the edges of the rectangular element are originally in these directions it remains rectangular. These directions are called the principal axes of the strain, and the stretches per unit length in these directions are called the principal extensions.

All the above is equally applicable to elastic solids and real fluids. The difference between them lies in the relation between stress and strain. But even here there is similarity up to a point, namely, where the stress is hydrostatic. If we compress a solid uniformly all round (which is most easily done by immersing it in a liquid and then compressing the liquid) and wait till it has come to rest, the result is a contraction in volume. Take the stress off, and the solid returns to its original dimensions. A liquid does precisely the same. In accordance with the molecular theory of matter we should expect a slight loss of energy in the process, leading to a rise of temperature, but experimentally this effect is too small to be detected. We may sum up these facts by saying that both solids and liquids are perfectly elastic for hydrostatic stress.

For distortional stress, on the other hand, there is a striking difference. Hang a weight on a wire; so long as the weight is not too great the wire settles down to a constant extension after a few vibrations. But pour treacle from a spoon; it continues to flow under its own weight till surface tension steps in and breaks the column. Twist a solid bar by a constant couple applied at the ends; it turns a certain amount and then stops. But suppose we have a liquid in a circular trough with a flat bottom, and that a circular horizontal plate is immersed in it. Turn the plate in its own plane by a constant couple, and it goes on turning at a constant rate till we take the couple off. The liquid in contact with the plate turns with the plate; that in contact with the bottom remains still. Thus we can produce as much distortion as we like by keeping the stress on long enough.

These facts can be expressed by separating the stress and strain each into two parts. Take the mean of the three principal stresses, and denote it by p. Subtract it from the three principal stresses and we are left with three quantities whose sum is zero. We can call these the distortional stresses. Similarly take the mean of the three principal extensions; it is equal to $\frac{1}{3}$ of the change of volume per unit volume. Subtract it from the principal extensions and we are left with a pure distortion. Then

8 EARTHQUAKES AND MOUNTAINS

both for solid and liquid we have a relation that the change in p is proportional to the volume expansion, and independent of the distortion. The change of p per unit volume expansion is called the bulk-modulus, or sometimes the incompressibility k; its reciprocal, the fractional change of volume per unit change of p, is called the compressibility.

For a solid it is found that the distortional stress applied is in proportion to the increase of distortion produced. The ratio is expressed in practice in terms of a further property of the material, called the *rigidity*. The meaning of this is seen most easily by considering a rectangular element deformed by forces in opposite faces. Suppose the stress, or force per unit area, over CD is S, and that as a result CD moves to C'D'. Then the ratio DD'/AD is the *shear*. It is proportional to the stress in CD, and the ratio is the rigidity μ. Stiff materials like iron have high rigidity; flexible ones like india-rubber have low rigidity.

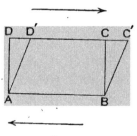

FIG. 2.

For a liquid it is found that under given stress the shear is not constant, but is proportional to the time. The ratio of the stress to the shear *per unit time* is called the *viscosity*.

For all liquids the results of this theory are found to agree with experiment. For most solids also they agree with experiment, provided the distortional stress is not too great. But when the distortion of a solid reaches a certain amount the strain ceases to be proportional to the stress, and many complications arise, which cannot yet be said to be fully understood. The general result is that at great distortional stresses every solid begins to partake to some extent of the properties of a liquid. Copper at moderate stresses satisfies the laws of elastic solids; but when the stress is great enough it can be made to flow through a tube like water, and afterwards shows no appreciable tendency to recover its shape. Other solids under sufficient stress will fracture; thenceforward the parts on the two sides of the fracture move independently, and one can acquire any amount of displacement with respect to the other. The distortional stress needed to produce such permanent changes in a solid may be called its *strength*. Just what determines it is uncertain. In part it is obviously characteristic of the material. According to the usual theory permanent change arises when the difference between the greatest and least of the principal stresses reaches a certain value.* This quantity is called the stress-difference. On this theory the value of the intermediate stress does not matter. On a recent theory due to Mieses, however, what matters is the sum

$$(p_1 - p_2)^2 + (p_2 - p_3)^2 + (p_3 - p_1)^2,$$

* This is equivalent to Coulomb's hypothesis of the maximum shear, published in 1776, and to the hypothesis of the maximum shear stress, as was shown by W. Hopkins, *Camb. Phil. Trans.*, 8, 456-466, 1849.

and G. I. Taylor and H. Quinney have shown that for some metals at any rate the beginning of permanent flow corresponds better with a constant value of this quantity than to a constant stress-difference. For geological purposes, however, the difference between the old and the new theories is probably unimportant. If p_2 is the intermediate stress, the Mieses function ranges only from $\frac{3}{2}(p_1 - p_3)^2$ to $2(p_1 - p_3)^2$ as p_2 varies. Thus even if the theory of Mieses is right the stress difference required to give failure varies only in the ratio 1 to $\frac{1}{2}\sqrt{3}$, or 0·866. This is well within the range of variation of strength of most of the materials that occur in geology, and for our purposes it is sufficiently accurate to say that failure is determined by the stress-difference. So far, however, the new theory has been tested only where yield takes place by continuous flow; it remains possible that the old one may be more accurate for fracture, but in any case the difference is not serious.* In this work I always mean by the strength the stress-difference needed to produce permanent deformation, whether it is flow or fracture.

The strength is obviously a property of the first importance to geology. Rock strata that must originally have been plane or nearly plane sheets are found in places to have been bent into deep arcs, in others to have been folded back to back like the bellows of a concertina, and in others again to have been definitely broken across, one side being displaced

* Cf. G. I. Taylor, *Proc. Roy. Soc.*, **145A**, 1-18, 1934.

by anything from inches to thousands of feet with respect to the other. We wish to know what stresses are required to produce these inelastic changes, and why the change is different in kind in different cases. A complete answer is not yet possible, but a good deal of relevant information is available, and we can at least exclude a sufficient number of theories, all more or less plausible at first sight, to be able to frame significant questions. We must not be disappointed if these cannot all be answered at once. In scientific work a question, once asked, often suggests at once the procedure for answering it; the really difficult part is to ask the right question.

So far we have seen that three physical properties are associated with the mechanical behaviour of solids, namely the bulk-modulus, the rigidity, and the strength. It is plain from elementary experience that these are different for different substances; granite has more rigidity than rubber, and more strength than shale. But all are influenced also by pressure, temperature, and, in the case of single crystals, by the direction of the principal stresses. It is well known that the refractive index of a crystal for light depends on the direction of the incident ray; similarly its elastic properties under tension depend on the direction of the tension with respect to the axes of the crystal, and if the stress difference becomes great enough to give permanent deformation the crystal shows a strong preference for sliding internally on planes determined by its structure rather than on others. Such materials really need more than two constants to express their purely elastic behaviour; the least symmetrical types of crystals actually need 21. In most geophysical problems, however, two constants are enough. The reason is that rocks are usually mixtures of crystals, which lie in all sorts of orientations, and even though any individual crystal may have special properties with reference to particular directions, the assemblage of millions of crystals has not. In addition some rocks are genuinely non-crystalline, like glass; how abundant such materials are is uncertain, but their elastic behaviour is expressible in terms of two constants. The directional properties of rocks, therefore, hardly concern us here, though they may possibly arise in stratified material, where we may reasonably expect different properties along and across the planes of sedimentation.

Temperature and pressure are certainly important. If we try to bend a piece of cast iron at ordinary temperatures, it either shows no visible effect or breaks right across; make it red hot and it bends permanently under a moderate stress. We notice that temperature here affects both the strength and the type of yield. In general rise of temperature seems to reduce the strength and to favour continuous distortion rather than fracture. Rise of pressure requires more elaborate technique before we can investigate its effects. In general it increases the strength, but also favours continuous distortion. In

SOLIDS AND LIQUIDS 13

the classical experiments of F. D. Adams and E. G. Coker, it was found that marble under a sufficient stress-difference at a high pressure could be made to flow continuously; this is impossible at low pressure, since it breaks. The combined effect of pressure and temperature may be expected to favour continuous deformation, since each has this effect separately; but the stress-difference that produces such deformation is raised by pressure and lowered by temperature. It is certain that in the earth both temperature and pressure increase with depth, so that permanent bending in rocks that would fracture under stress under surface conditions is evidence that they have been deeply buried; but there are difficulties about experimenting with high temperatures and pressures simultaneously, and the influence of depth on the strength can hardly be estimated from laboratory determinations. We can proceed best by seeing what the earth itself has to tell us.

Temperature and pressure also affect the bulk-modulus and the rigidity. These are raised by pressure but lowered by temperature in most cases, though a few exceptions are known. Here again there are difficulties about the application of the experimental results. Pressures up to about 30,000 atmospheres have been obtained in the laboratory. In the earth's crust the pressure increases with depth at the rate of about 1 atmosphere in 4 metres, so that a pressure of 30,000 atmospheres would correspond to a depth of 120 km. The temperature rises with depth at an average rate of about 30° C. per kilometre, so that a temperature of 1800°, which would melt nearly all rocks, would be reached at a depth of 60 km. if the rate of rise was maintained. It appears therefore that the ranges of variation of pressure and temperature attainable in the laboratory are enough to cover those involved in geological processes; on the other hand, they do not cover more than a fraction of the whole radius of the earth (6370 km.), and any extrapolation to great depths would be dangerous. Nevertheless we cannot approach a full understanding of what we observe at the surface without considering what is going on at great depths, and our laboratory knowledge has to be supplemented from observation of the earth itself.

The foregoing account of the relation between stress and strain might suggest that the distortions are proportional to the differences between the principal stresses right up to the breaking point. This is still an over-simplification, though there are so many problems (chiefly in engineering) where accurate proportionality holds that we must regard it as having a high degree of generality. In any case it seems to be accurately true of the immediate results. When a bar is bent by hanging different loads on it, the immediate displacement seems to be accurately proportional to the load. For many substances, so long as the load is not too great, nothing further happens as long as the load is left on; and when the load is taken off the bar returns to its original form. But

SOLIDS AND LIQUIDS 15

sometimes when the load is left on the displacement continues to increase for a long time; and the rate of increase is not necessarily proportional to the load. In some substances, such as pitch, it may increase indefinitely if the load is left on, till all resemblance to the original shape has disappeared. Such substances may be called plastic or elasticoviscous. But with other materials the displacement does not increase beyond a certain amount, however long the load is kept on. When the load is removed, the bar springs back by an amount equal to the original elastic displacement. But then it goes on recovering, and finally, possibly after a long time, it returns to its original form as accurately as we can measure. Such behaviour is called *elastic afterworking*. In a sense it involves no permanent deformation, but it is a true imperfection of elasticity. The weight while it is left on is doing work on the bar, which is converted into heat by some kind of internal resistance, and this energy is not restored during the subsequent return.

It sometimes happens that a load once applied does not give fracture immediately, but that when the load is left on the bar breaks after a certain time when the displacement has reached a certain stage.* It is therefore sometimes necessary to distinguish between the stress-difference that produces fracture immediately and that which produces it when it is

* D. W. Phillips, *Trans. Inst. Mining Engineers*, **80**, 212-242, 1931; **82**, 432-450, 1932.

maintained for a long time.. This fact has a geological application. We often see rock strata bent into an arch without fracture; but sometimes the top of the arch is broken. The stress has here produced a considerable amount of continuous flow, but at some stage after flow has begun it has given place to fracture.

These types of imperfect elasticity represent a compromise between the behaviour of elastic solids and liquids. If we apply a stress to an elastic solid gradually,* so as to avoid setting up oscillations, the whole of the work done by the stress reappears as potential energy of the strain. If the stress is again taken off gradually, the solid does just as much work against the stress during recovery, as the stress did on the solid in the first state; there is complete recovery of form and no loss of energy. But if we do the same thing to a liquid the stress is doing work on the liquid as long as it flows, and this work is converted into heat; there is no perceptible tendency to recover when the stress is removed. To understand what happens in intermediate types of substance really presupposes that we understand liquids, and it is unfortunate that nobody has any real idea about why liquids exist. We understand crystals in terms of the arrangement of their constituent atoms in a repeating pattern; they are stable systems, and their elastic properties can be quantitatively explained. We also understand gases in terms of the kinetic theory, according

* If we put it on suddenly and leave go, oscillations are set up, and part of the work done goes into the kinetic energy of the oscillations.

SOLIDS AND LIQUIDS

to which the gas consists of molecules moving independently except when in collision. But we do not understand why there is a range of temperature such that the crystalline and gaseous states are both unstable, and are replaced by the curious compromise that we call the liquid state. Liquids resemble solids much more closely than gases in their densities and their bulk-moduli, and examination of them by X-rays indicates that a considerable part of the crystal structure persists. The atomic pattern is more disturbed than in a crystal, but clots of many molecules in about the same relative positions as in a crystal are found at any moment. The difference is that they are not permanent. The thermal agitation is continually shaking molecules off the outside of the clots, so that there are also freely moving molecules as in a gas, which in turn often combine with one another or with existing clots. Thus the clots, while they exist at any moment, have no permanent individuality. Such a constitution gives a qualitative, and in some respects a quantitative, representation of the properties of liquids. The mechanism of viscosity, which is what concerns us now, is as follows. Suppose the liquid is between two parallel plane solid surfaces, and that we apply to one of these a stress in its own plane. The clots in contact with it are moved along with the wall, being held together temporarily by the interatomic forces. But molecules on the other side of these clots become detached by thermal agitation, and in time become attached to more remote clots, to which they convey the momentum acquired from the boundary. Thus the momentum supplied at the boundary gradually becomes dispersed through the whole. But the other wall remains at rest, and momentum reaching the clots in contact with it is removed. Thus all points of the liquid are set in motion except at the fixed wall, and momentum is continually being transferred through the body of the liquid from the places of higher velocity towards the fixed wall. The essential difference between the liquid and the crystal is that in the latter the momentum is transferred directly through it by the atomic forces. Once the lattice is distorted, even if the distortion is not increasing with the time, these forces are altered in direction and produce a transfer of momentum. To maintain a steady distortion in a crystal therefore requires forces to remove the transferred momentum, and therefore stress over the boundaries. In a liquid, once the clots are at rest, there is nothing to make more of the free molecules move forwards rather than backwards, and transport of momentum stops.

Such a theory explains also the variation of viscosity with temperature and pressure. If the temperature is raised, there is more violent thermal agitation, and the free molecules become more numerous at the expense of the clots. But within the clots momentum is transmitted by elasticity, and an elastic wave travels much faster than a free molecule. Thus temperature reduces the rate of redistribution of

SOLIDS AND LIQUIDS 19

momentum; in other words, the viscosity decreases as the temperature rises. A quantitative theory of this aspect of the problem has been given by Andrade,* who obtains a good agreement for many liquids. Pressure reduces the volume, the free molecules have not so far to travel before striking clots, and the viscosity rises. This is in qualitative agreement with the facts as ascertained by Bridgman,† but the quantitative theory is still incomplete.

But our distinction between quasi-solid clots and quasi-gaseous free molecules may be too sharp. Imagine a nearly level surface, with a number of slight elevations and depressions. In some of the latter marbles rest. Such marbles are in stable equilibrium, and may be likened to the atoms in a solid. Apply a small force to a marble, and it is slightly displaced, returning to its original position when the force is removed. But if we apply a large enough force the marble may be pushed over an elevation into a neighbouring depression, giving a permanent displacement, not recoverable when the force is removed. With a little more force the marble may be unable to come to rest anywhere, and we have an analogue of deformation increasing with the time. This is, however, too simple to fit the facts. It would suggest that the stress needed to produce permanent deformation in a crystal would have to be enough to move each atom in a layer into the

* *Phil. Mag.*, **17**, 497-511, 698-732, 1934.
† *The Physics of High Pressure*, 1931.

position of the next; a sodium chloride crystal, for instance, should not break until its square faces have become parallelograms with angles of 45° and 135°. This would mean that the strength should be of the same order of magnitude as the rigidity. Such strength is attained by rubber, but by no crystal.

Our model, however, neglects the effect of heat. To imitate this the marbles must be supposed not to be at rest, but to be oscillating about their positions of equilibrium with different amplitudes. If the amplitude of the oscillation is great enough a marble may be able to wander freely from one position of equilibrium to another; there are no permanent positions of equilibrium, and we have an analogue of a fluid. The phenomenon of melting is probably of this nature. But suppose that the agitation does not reach this amount; it nevertheless aids permanent displacement when a force is applied. If the same force is applied to every marble, those with the greatest amplitudes may be pushed over, while others remain in their original pits. The motion of the former is resisted by friction, of the latter by gravity. Thus if we consider the system as a whole, both elastic and frictional resistances may occur. Returning to the solid, we see that some atoms may behave as if free, while their immediate neighbours stay near their permanent positions. But the energy of the thermal motions is being continually interchanged between one atom and another, so that we cannot regard one atom as permanently free and another as permanently fixed. The question becomes very complicated, and little progress has been made towards a solution. It appears that pure single crystals of metals and at least some compounds behave as if they possessed no strength; they acquire strength only when impurities or irregularities of the internal arrangement partly destroy the regular crystal structure.* We have to remember at the same time that the force in a crystal tending to keep an atom near a fixed position is not an external one like gravity, but arises from the attractions and repulsions of its neighbours, so that if one atom is displaced and loses much of its thermal energy, it will pass this on to the surrounding ones, some of which will become more easily displaced.

These considerations suggest that when a solid is undergoing flow we must regard the distortional stress as composed of two parts, one proportional to the strain and resisted by rigidity, the other proportional to the rate of increase of strain and resisted by viscosity. Such a theory was first suggested by Maxwell, and developed by J. G. Butcher, Sir G. H. Darwin, and more recently by Mieses and Reuss. When strength is finite the viscosity must be regarded as infinite so long as the stress-difference does not reach the strength. When there is no strength the substance may be called liquevitreous or elasticoviscous.

Is there then any essential difference between an elasticoviscous solid and a liquid? The foregoing account would suggest that the criterion should be

* G. I. Taylor, *Proc. Roy. Soc.*, **145A**, 362-415, 1934.

that the liquid possesses no rigidity, while the solid does, and I have adopted this test in previous work. The difficulty, however, is in applying this test experimentally. The interatomic forces in a liquid must be expected to be much the same as in a solid, and in both the rigidity would be expected to be comparable with the bulk-modulus. In water we should expect a rigidity of the order of 10^{10} dynes per square centimetre. But the viscosity of water is only 10^{-2} in c.g.s. units. If then a distortional stress persists for as long as 10^{-12} second the viscous displacement would equal the elastic part, and therefore, even if water has rigidity, there is no practical possibility of detecting its effects except under forces varying in what seem to be impracticably short periods. It is only when the viscosity is high that the effects of rigidity would be detectable.

Now single crystals do not occur naturally. In most solids there are interfaces between crystals, and at these the crystal form is distorted; the atoms have to divide their allegiance between the adjoining crystals. In glasses we cannot detect any crystalline structure by ocular or microscopical examination, though traces of it can be found in very small regions by X-rays. At low temperatures these conditions make for strength. The chief source of weakness in a crystal is that when one atom becomes displaced out of a position of equilibrium it increases the probability that its neighbours along a certain plane will be displaced too, and fracture or flow will take

SOLIDS AND LIQUIDS

place along crystal planes. But in a crystal aggregate, an impure crystal, or a glass, there are no such planes for the body as a whole, and the principal source of weakness is removed. If a single metal crystal is deformed repeatedly, it partly loses its crystalline structure, as X-ray observations show, and at the same time the strength increases. On the other hand, at a high temperature an imperfectly crystalline substance loses much of its strength, for its atoms have different ranges of oscillation before they become able to change their mean positions, and some at any rate are in this sense less stable than in a crystal, and inelastic deformation can set in at a lower stress. Thus we should expect that far below the melting-point non-crystalline substances should be stronger than crystals of the same material, but that near the melting-point this relation will be reversed. This is true; at ordinary temperatures silica, with a melting-point of about $1700°$, is much stronger as a glass than as a crystal, but sulphur, with a melting-point of $445°$, has no strength at room temperature in the amorphous state known as "plastic sulphur". Glasses in this state are liable to crystallize, or "devitrefy," though other glasses may retain their state for thousands or millions of years.

If a glass or an imperfect crystal is acted on by a distortional stress, it at once acquires an elastic displacement. If the stress is maintained, atoms are displaced inelastically and the strain increases, in some cases indefinitely. But if only a fraction of the

atoms are capable of inelastic displacements, the strain will grow only until the stress is borne entirely by the elasticity of the rest of the structure. If the stress is removed, the less stable atoms do not at once spring back; the immediate recovery of strain will be equal to the original elastic strain, but these atoms will be gradually pulled back to their original positions by the elasticity of the rest of the structure. Thus the mechanism explains elastic afterworking.*
Now if such imperfections of crystal structure are the cause of elastic afterworking we should expect it to be absent from perfect crystals, and it is. Crystals that have a finite strength at all are found to be free from elastic afterworking.

It appears therefore that known mechanisms can account qualitatively for the characteristic mechanical properties of crystals, liquids, and solids showing plasticity and elastic afterworking. The theory has not yet been worked out in full quantitatively, and is certainly very difficult. But it is important for geophysical purposes to have all the possible alternatives stated, especially as it has sometimes happened that properties have been described as mutually inconsistent that quite certainly are not. Thus some authors define a solid explicitly as a crystal, others as a substance possessing a finite strength, and arguments are then based on the postulate that nothing but a crystal can have a finite strength. This is simply untrue; pure single metal crystals have no strength, and

* Jeffreys, *Proc. Roy. Soc.*, **83A**, 283-297, 1932.

the strongest material known is silica glass. Again, any sort of imperfection of elasticity is sometimes thought to imply that the substance concerned will be deformed to an indefinitely large extent by any force, however small, provided that this is kept on long enough; this is not true, because it confuses the characteristically different properties of plasticity and elastic afterworking. Unfortunately we know very little about the effect of high pressure on all these properties, especially when the temperature is also high, and we cannot therefore make theoretical predictions about the constitution of the earth's interior from present data. All we can do is to call attention to the variety of the combinations of properties that do occur; for their geophysical application we must ask the earth itself and trust its answer, not assuming that one property necessarily implies another when it is known that they do not always go together even under laboratory conditions.

The practical definition that will be adopted in the following pages is that a substance is regarded as solid if it can be observed to transmit distortional waves. The alternative definition, that a solid is a substance with a finite strength, is apparently consistent with this one, but cannot be applied in such detail; we can say of every part of the earth whether it does or does not satisfy the first definition, but the second criterion can only be applied in general terms.

CHAPTER II

EARTHQUAKES

"As the Book of Verses indicates, 'The person who patiently awaits a sign from the clouds for many years, and fails to notice the earthquake at his feet, is devoid of intellect'."
—*The Wallet of Kai Lung*, 16.

THE study of earthquakes really consists of three parts: the causes leading up to the earthquake, the motion produced by it in the earth, and the effects of this motion on humanity. To begin with a very abbreviated statement, we may say that an earthquake is the result of an internal stress leading up to a fracture. This fracture generates waves, which travel through the earth. These are strongest in the neighbourhood of the fracture, and often produce a motion of the ground of sufficient intensity to be felt, or in extreme cases to damage buildings. The last phase of the problem attracts the most popular attention, even in a country like England where destructive earthquakes are practically unknown; but it depends on the nature of the waves, which also provide more detailed information about the structure of the earth than is provided by any other line of investigation. They are found to show that the earth is solid, in the sense defined at the end of the last

EARTHQUAKES

chapter, to a depth of about half the radius; the material at greater depths appears to be liquid.

The nature of earthquake waves may be illustrated by comparison with the motion produced in air by an explosion. An explosion gives a sudden local increase in pressure. This extra pressure acts on the surrounding air in two ways: it compresses it and at the same time pushes it outwards. But the air driven outwards then presses on other air, and so the disturbance spreads. If instead of an explosion we have a continuous vibration, as in an engine whistle, the motion in the air is a train of waves. At every point in the air the medium is vibrating in and out, and at the same time the pressure is varying; the pressure is high where the air is moving outwards from the source of sound, low where it is moving inwards. The waves travel outwards with a characteristic velocity, the velocity of sound in air. In the case of an explosion, it is found that at distance r from the explosion the air remains at rest, with no change of pressure, until time r/c after the explosion, where c is the velocity of sound. At this time the air is suddenly displaced outwards and the pressure rises. It is necessary for some purposes to distinguish between a wave train, which is a long train of waves, and a pulse, which is a single displacement such as an explosion produces, but the velocity is the same for both.

In water similar waves and pulses can be produced. But another type of wave is conspicuous in water when acted on by a wind or when a solid is thrown into it. The surface of the water is then thrown into waves travelling away from the source of disturbance. On deep water these have the property that their size is greatest at the surface; at a depth of a wave-length or more the displacement at any point is a small fraction of what it is at the surface. Such waves are called surface waves.

In an elastic solid two different types of waves exist, besides surface waves. If an outward force is applied over a sphere inside the solid, it sends out a wave of the same character as in a fluid. But a distortional stress also produces a wave. If we apply a shear stress over all points in the plane $x = 0$, tending to pull them in the direction of increasing y, they are displaced elastically, but matter at some distance away remains undisturbed for a time. The difference of displacement gives a shear, which implies by the laws of elasticity a stress on this more distant matter, which is therefore set in motion. Thus a wave of distortion travels out, every particle being displaced transversely as the wave passes. We have to distinguish carefully in seismology between these two types of wave, because they differ both in velocity and in the motion produced. In the former type, as the wave advances each particle of the solid is displaced in the direction of travel of the wave; in the latter the displacement is at right angles to the direction of travel. The former type of wave is therefore longitudinal, like a sound wave; but the latter is transverse, as in a light wave. The transverse waves resemble light in showing polarization. If the waves are travelling along the axis of x, displacements in the directions of the y and z axes are completely independent, in the sense that either can exist without the other. If they both exist they are transmitted with the same velocity, but differences arise on reflexion and refraction, again as in the case of light.

The velocity of the longitudinal wave is $\{(k + \frac{4}{3}\mu)/\rho\}^{\frac{1}{2}}$, where k is the bulk-modulus, μ the rigidity, and ρ the density; that of transverse waves is $(\mu/\rho)^{\frac{1}{2}}$. The former is the greater, so that if the original disturbance is partly outwards and partly transverse, the longitudinal wave will be the first to arrive at any other point. For this reason the longitudinal wave is often called the primary and the transverse one the secondary, and these words are conveniently abbreviated into the standard notation of P and S. As the late Professor H. H. Turner remarked, the letters P and S may also be regarded as descriptive of the character of the motion; P gives a push (or sometimes a pull) and S a shake.

Surface waves also can occur in a solid. In general features they resemble surface waves on water, except that they are controlled by elasticity, whereas the surface waves on water are controlled by gravity, which is negligible in earthquake waves. They are of two types. In one, predicted theoretically by the late Lord Rayleigh, the displacement of the surface is partly vertical and partly in the direction of propagation. It can exist on a uniform solid; the velocity of travel is about 0·92 of that of an S wave. The other type was discovered by Professor A. E. H. Love. It is possible only if the material is not uniform. There must be a superficial layer resting on another layer, the velocity of distortional waves being less in the upper layer than in the lower. The displacement in these waves is wholly horizontal and at right angles to the direction of propagation. They show the phenomenon of dispersion; that is, the velocity of travel depends on the wave-length, which is not true for P and S waves. More strictly, the velocity depends on the ratio of the wave-length to the thickness of the layer, and therefore if the velocity is found by observation for a number of wave-lengths we have a means of finding the thickness of the layer. The velocity of Rayleigh waves similarly depends on the wave-length if the crust is not uniform, and can be made to provide similar information, but the theory is more complicated.

The waves P and S, which travel throughout the material, may be called " bodily " waves to distinguish them from the two types of surface waves. They were predicted theoretically by Poisson in 1829, while Rayleigh's discussion appeared in 1887. But it was not until 1900 that the three types were distinguished in actual records of earthquakes by R. D. Oldham.. This delay was chiefly owing to the fact that instrumental seismology is a comparatively new subject.

EARTHQUAKES 31

The fundamental principle of most seismographs is that of a pendulum. Suppose that a pendulum is hanging at rest, and that the earth's surface is suddenly displaced horizontally, carrying with it the shaft that the pendulum hangs from. The bob of the pendulum does not begin to move at once ; the immediate result is a change in the inclination of the string, which is proportional to the displacement of the ground if it takes place sufficiently suddenly. Thus a simple pendulum, if we can record its changes of inclination, is suitable for recording the motion of the ground. But actually the simple pendulum suffers from certain defects, which make it unsuitable until modifications have been introduced. The actual movement in an earthquake is an oscillation with a period of some seconds, while a pendulum has to be inconveniently long if it is to have a period of more than one or two seconds. Now if the support moves in a period long compared with the natural period of the pendulum, the bob has time to follow the support, and the inclination, which is the only thing we can measure, varies very little. To get a good record we need an instrument with a period longer than that of the waves in the earth, which may be about 7 seconds for P and longer for other waves. But a simple pendulum, to have a period of 10 seconds, say, would have to be 25 metres long ! We therefore need first of all a device for lengthening the period. In many seismographs, especially the Galitzin and Milne-Shaw machines, this is done by making the pendulum swing about a nearly vertical support, to which it is hinged like a farmyard gate. The support must not be exactly vertical, of course, otherwise the instrument could remain at rest in any position. But with a small inclination to the vertical the period can be made as long as we like: natural periods of about 10 seconds are obtained in actual instruments with booms about 20 centimetres long. With such a device we need separate instruments to record the motion of the ground to the north and to the east. Thus if the boom is hanging to the east its direction is affected by displacements of the ground to the north or south, but not by displacements to east or west. If we wish to record both the easterly and northerly components of the displacement we need two pendulums. In the instruments of Wiechert the pendulum is abandoned. These instruments consist of a heavy mass on a vertical rod, hinged at the bottom. This would simply fall over with a heavy crash if it was left to itself (in the three types of Wiechert machine the masses are respectively 80 kilograms, a ton, and 17 tons) but it is kept in position by four springs. When the ground moves the springs yield elastically and displacements in both directions are recorded. In the Wood-Anderson instrument a cylindrical mass is fixed eccentrically to a vertical wire; when the earth moves the wire is twisted, and its tendency to untwist itself provides the restoring force.

Other instruments have been designed to record

EARTHQUAKES

the vertical movement, but only a fraction of the observing stations possess them. The difficulty of obtaining a long period is greater than for horizontal instruments, but it can be overcome, and for recording some waves the vertical instruments are the most valuable.

Instrument design would be a fairly simple matter if we had to record only the first movement. The motion due to an earthquake involves a large number of identifiable phases, and each one has to be recognized by the motion it produces in the instrument. Difficulties will arise in identifying any phase if the motion due to the previous ones has not died down. We try therefore to ensure that the motion due to any phase will have disappeared before the next arrives. This is done by introducing damping. The motion of the mass is resisted by a force proportional to its velocity, which may be produced either by fluid viscosity or electrically. In the former case the motion may be made to drive a small vane through oil, or air through a narrow orifice. In the latter, the boom carries a small magnet, which moves near a coil of wire. When the magnet moves it induces an electric current in the wire. The resistance in the wire puts the current out of phase with the motion of the magnet, and the force of the current on the magnet gives the resistance required. If there was no damping and the instrument was once set in motion, it would go on vibrating for ever: it turns out also that if damping is too heavy the displacement disappears unnecessarily slowly, so that there is an optimum degree of damping. The usual arrangement is that each swing is about $\frac{1}{20}$ of the one immediately before it.

In this way we make the displacement of the bob, whatever the form of the instrument, resemble that of the ground as closely as is physically possible. It is usually still too small to be visible: it must be magnified, and a permanent record must be made. In the Wiechert type of instrument the magnification is done by a system of levers, and a fine point is ultimately made to move about 200 times as far as the heavy mass. A cylindrical drum carries a sheet of smoked paper and rotates at a uniform rate, the fine point being in contact with the paper. As the point moves it scrapes the soot off the paper; meanwhile the drum rotates, carrying the paper in a direction perpendicular to the motion of the point. The result is a white line on a dark ground, which is effectively a graph of the displacement of the heavy mass with respect to the time. In the Milne-Shaw and Wood-Anderson instruments the paper used is photographic, and the moving parts are made to reflect a small spot of light upon it. When the paper is developed we get a dark line on a white ground. This arrangement has the advantage that solid friction is absent or nearly so: in the smoked paper machines friction in the levers is considerable, and can be overcome only by making the instrument very heavy. In the machines designed to record very small move-

EARTHQUAKES 35

ments, with magnifications of about 2000, the mass is usually about 20 tons. An independent mechanism connected with the observatory clock is made to mark exact minutes on the record : this is necessary because in practice the drum never rotates as uniformly as we should like, and times have to be read to a second or less.

There are now some hundreds of seismological stations in various parts of the world.

Oldham's study showed that the record of an earthquake at any station showed a sudden commencement, which can be identified with P : this was followed by a train of waves, but after some time there was another sharp movement, usually larger than P ; this was identified with S. The times of arrival of these two phases at different distances were found to give two smooth curves, which gave the first empirical knowledge of the times of transmission of elastic waves through the earth. S was followed by a series of smooth waves, with amplitudes larger than that of S. They increased in size up to a maximum and then slowly died down. These were considered at first to be the Rayleigh waves. Their comparatively large size can be understood from their mode of spreading. In a uniform solid the bodily waves spread out symmetrically, downwards as well as horizontally. The energy transmitted must be the same at all distances, and therefore the energy crossing unit area of surface must vary inversely as the square of the distance. But the surface waves do not penetrate more deeply as they advance, and the energy per unit surface at small depths is inversely proportional only to the distance. Hence at great distances, and for observations at ground level, the surface waves may be expected to be larger than the bodily ones.

This interpretation of the surface waves was, however, too simple to cover the facts. Rayleigh waves involve displacement in only two directions, vertical and in the direction of travel. They should give no horizontal displacement at right angles to the direction of travel. Actual records showed strong movement in this direction. This was explained by Love in his theory of surface waves when there is a superficial layer. The two types of surface wave have been clearly separated by Stoneley. Earthquakes were found such that the waves from them arrived at Kew Observatory from due east; then the Rayleigh waves were shown on the vertical and easterly records and the Love type on the northerly. The two types begin at different times and the movements in any given minute have different periods.

We should expect that the time needed by surface waves to travel a given distance would be proportional to the distance measured over the surface, and this is found to be true. But if the earth was uniform the time of travel of bodily waves would be proportional to the distance measured in a straight line from the origin to the observing station, and this distance increases less rapidly than the arc. Consequently it was expected that the times of travel of the bodily waves, when plotted against the distance measured over the surface, would give a curve with a downward curvature. This actually happened; but it was found that the time was not even proportional to the chord. If we estimate a mean velocity as the length of the chord divided by the time of travel, this velocity is found to be greater the greater the distance. The only possible interpretation of this result is that the velocity increases with depth. Further, given the times of transmission to different distances, we can actually calculate the velocity at any depth reached; the theory is somewhat involved, because the rays are not straight but curved upwards when the velocity increases with depth, but the solution is quite definite.

The usual practice in seismology is to measure a distance as an angle subtended at the centre of the earth, so that a distance of 90° means that the origin and the station are a quadrant, or 10,000 kilometres, apart. At a distance of about 105° the P and S waves seem to fade out. Waves reaching this distance must have penetrated to a depth of about 3000 kilometres, nearly half the radius of the earth, and their velocities at this depth are about 1·7 times those at high levels. Oldham found that a new wave, apparently of P type, emerged at distances over 140° up to the antipodes. If such a wave travelled through the inner region with the same velocity as it had at a depth of 3000 kilometres, it should reach the antipodes about 16 minutes after the shock: the actual time is 20 minutes. The only interpretation of these waves is that their velocity in the central region is about 9 km. per second; the velocity at a depth of 3000 km. is 13 km./sec., the reduction accounting for the observed delay. Waves of S type through the central region cannot be identified.* This of course suggests strongly that the central region is incapable of transmitting distortional waves, and therefore is liquid.

Later work has confirmed this suggestion. It is found that we must regard the earth as composed of a solid shell, reaching nearly half-way to the centre, and resting on a liquid core capable of transmitting P but not S waves. When an S wave reaches it, it is partly transmitted as a P wave and partly reflected. These P waves in the core travel on until they strike its boundary again, when they are again partly reflected and partly transmitted as a pair of P and S waves, which go on in the shell till they reach the surface. Such waves are recognizable on the records and give further information about velocities in the core. Gutenberg has found it possible to calculate the times of travel of waves reflected both at the outside and the inner surface of the core, and movements are found on the records at practically the calculated times. These waves are of special interest, because they show that the boundary of the core is not merely a gradual transition, but a sharp interface.

* Mr. L. Bastings (*Nature*, **134**, 216, 1934) claims to have detected such waves now, but I think that his interpretation raises other difficulties and is not the only one possible (*Nature*, **134**, 324, 1934).

EARTHQUAKES

If the transition was gradual, the reflected waves would be so spread out in time as to be unrecognizable. The nature of the material of the core cannot be determined by purely seismological methods, but other information is given by gravity, and indicates that the fall of velocity is associated with a rise of density, and that the core probably consists of a heavy metal in the liquid state. It appears that the primitive earth consisted of two immiscible materials, and that the denser sank to the centre and has remained liquid ever since, in spite of the changes that have gone on outside it.

We have seen that the existence of Love waves implies a rather sudden change of velocity of distortional waves at a small depth. Now such a change of materials was inferred by Suess on purely geological grounds. There are many types of igneous rocks known, which have been in some way melted underground and have arisen to be poured out over the surface or intruded into rocks near the surface that have become exposed to view by denudation. The two most abundant types may be described roughly as granite and basalt, though rocks in each class vary appreciably in chemical composition and state of crystallization. Further, one of the most abundant sedimentary rocks, sandstone, consists principally of quartz grains that have been derived from granite by chemical weathering. Granite is found all over the continents, and the abundance of sandstone indicates that there has been even more of it in the past. But in oceanic islands granite is distinctly rare. Thus for the Pacific Ocean, P. Marshall* records some granite on Viti Levu (Fiji), but none elsewhere. These islands are mostly made of basalt, with the intermediate rock andesite, and tending towards rocks more remote from granite in silica content than basalt is. There is a strong suggestion that basalt is a universal rock, whereas granite represents an upper layer characteristic of the continents. The first seismological recognition of such a structure in Europe was due to Professor A. Mohorovičić. He noticed in a Balkan earthquake that the P movement consisted of two phases. At short distances there is a large sharp movement, followed after some seconds by a larger S movement that appears to have followed much the same path. At distances of more than about 150 km. both these movements are still conspicuous, but the large P movement is preceded by a smaller one; the greater the distance the greater is the interval between the two P movements. The S movement is duplicated in the same way. It is clear that the smaller wave in each case travels faster; but when we compare the times of transmission we find that it seems to have started later. For all four waves the time of arrival gives a straight line when plotted against the distance; in other words, the times are closely equal to expressions of the form $a + \Delta/v$, where Δ is the distance, v the velocity, and a a constant, which we may call the apparent time of starting.

* *Handb. d. Regionalen Geologie*, Bd. 7/2, Heft 9.

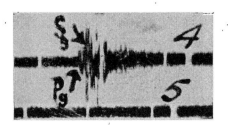

WEST BROMWICH ($\Delta = 63$ KM.), RECORD OF THE HEREFORD EARTHQUAKE OF 1926 AUGUST 14, SHOWING ONLY Pg AND Sg, WITH A TRAIN OF OSCILLATIONS FOLLOWING THE LATTER. A CHARACTERISTIC RECORD AT A VERY SHORT DISTANCE.

EARTHQUAKES

That is to say, the times are the same as they would be if the wave had started at time a and travelled the distance with velocity v. But not only v, but also a, has a different value for each wave. The smaller P wave usually has a value of a exceeding by 6 or 7 seconds the value for the larger one. Now on the supposition of a layered structure these facts are immediately explained. Suppose that we have an upper layer resting on a lower one, the velocities of P and S being less in the upper layer than in the lower one. If an earthquake happens in the upper layer it sends out P and S waves in that layer, which travel directly to the observing stations with their appropriate velocities. But the P and S waves will strike on the interface between the two layers, and will there be refracted into the lower layer, the ray in each case being bent away from the normal. Thus refraction will give also P and S motion travelling horizontally in the lower layer: this can be refracted up again at any distance. The apparent velocity is the velocity in the lower layer: but such an indirect wave wastes a certain amount of time in travelling down and up in the upper layer, and this loss of time accounts for the apparent delay in starting. At short distances the direct waves arrive sooner than the indirect ones, for much the same reason that up to distances of five miles or so we shall usually arrive sooner by using a bicycle instead of a car, because it takes time to get the car out of the garage.

Similar results have been obtained in a number

of other earthquakes, not only in Europe, but in California and Japan. The general picture, however, tends to become more complicated. Earthquakes in which the upper layer waves are recognizable are technically known as "near earthquakes", because at a distance of about 800 km. these waves lose their distinctive character.

The identification of the layers depends on the comparison of the observed velocities with laboratory determinations of the elastic properties of rocks. Given the velocities of the P wave and the corresponding S wave we can calculate k/ρ, the ratio of the bulk-modulus to the density. But the density and the bulk-modulus can be found directly in the laboratory. It is found that the velocities of P and S in the upper layer are consistent with this layer being granite. It was expected that the lower layer would be basalt, but it has turned out that this is not the case. The velocities are definitely too high to fit basalt or any of its ordinary modifications. They do fit olivine, a mineral composed of a mixture of magnesium and ferrous orthosilicates. It is a common mineral in basalt and other rocks, but seldom forms the chief constituent of a rock accessible at the surface. Such a rock exists, and is known as dunite, but is rare.

The recognition of the high bulk-modulus of the lower layer makes our natural theory of the origin of basalt untenable. Basalt contains a considerable percentage of calcium, sodium, and aluminium, all of which are absent from olivine. It therefore cannot

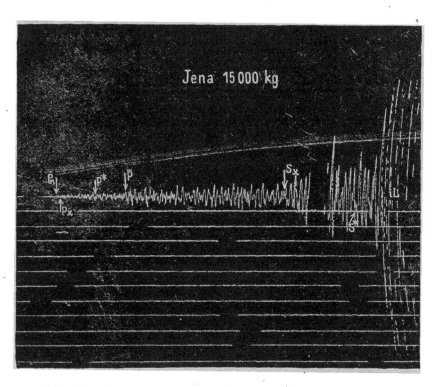

JENA ($\Delta = 481$ KM.), RECORD OF THE SCHWADORF EARTHQUAKE OF 1927 OCTOBER 8; SHOWING THE OTHER PHASES THAT EMERGE AT INTERMEDIATE DISTANCES

apparently have come from the lower layer. (We reserve at present the possibility that the lower layer may consist of different minerals from basalt, but with the same general composition.) We therefore need an alternative explanation of the origin of basalt.

Laboratory investigation of the properties of rocks has led to the recognition of a number of general rules, which can be applied even when the exact composition is doubtful. Most rocks are chemically silicates, made up of silica in combination with different amounts of various metallic oxides. Silica is regarded as an acid-forming oxide, just as sulphur trioxide is considered an acid-forming oxide in sulphuric acid and the sulphates. It occurs by itself as quartz. The metallic oxides are on the whole denser than silica, and as a general rule the greater the amount of metal associated with a given amount of silica, the greater is the density of the mineral or rock. In addition, increase of the metal increases the bulk-modulus even more than the density, so that the velocities of transmission of elastic waves also increase. Incidentally quartz has a high thermal conductivity, and on the whole high silica content is associated with high conductivity. We have therefore the general rules that as the silica content of rocks decreases, the density and the velocities of elastic waves rise, and the thermal conductivity falls. There are some exceptions to these rules, but what matters to us is that the commoner rocks and minerals follow them.

Now if the igneous rocks found at the surface can be considered as derived from layers within the crust we shall expect the denser rocks to lie below the lighter ones. The density of granite is about 2·7, that of olivine about 3·3 ; basalt, with a density of about 3·0, would then be expected to lie between the granitic and olivine layers. It was therefore expected that seismology would reveal an intermediate layer with the properties of basalt. Recent work has shown that an intermediate layer exists. The first evidence for it was obtained by Professor V. Conrad, who detected in an Austrian earthquake a wave of P type with a velocity of 6·2 km./sec. But this is too low for crystalline basalt ; the velocity in this would be about 7·0 km./sec. Adams and Gibson [*] find that tachylyte, which is basalt in a glassy state, would give a velocity of 6·4 km./sec. So far as these facts go, we should be entitled to suppose that the intermediate layer is basalt only if that basalt is glassy and not crystalline. There are some difficulties about such an interpretation, and personally I have always preferred to speak of the " intermediate layer " without specifying its composition, unless the composition is relevant to the question immediately under discussion ; fortunately in many of our problems the composition is not important, on account of the intimate relation that holds between velocity and density. But further work has led to more complicated results still. The

[*] L. H. Adams and R. E. Gibson, *Proc. Nat. Acad. Sci.*, **15**, 713-724, 1929.

EARTHQUAKES 45

velocity of the wave in the upper layer, which is now usually known as P_g, is about 5·6 km./sec.; that in the lower layer is about 7·7 km./sec.; but there is evidence for the existence of P waves in three distinct intermediate layers, the most probable values of their velocities being 6·3, 6·5, and 7·1 km./sec. The S waves corresponding to the first two are also known, and have velocities of 3·6 and 3·74 km./sec. We could identify the third intermediate layer with crystalline basalt, but the evidence for its existence is less satisfactory than for the others, since it has been recognized in only one earthquake and the corresponding S wave has not been found. A further difficulty is that the waves have not all been recognized in any single shock.

We cannot therefore say definitely where the basalt comes from, except that it represents one of the intermediate layers.

The constant terms in the times of arrival of the various waves are affected, as we have remarked, by the time lost in travelling down to and up from their characteristic layers. Consequently they give information about the thicknesses of the layers. The differences are small, the whole time lost by P_n (the pulse of speed 7·7 km./sec.) in comparison with P_g being about 6 or 7 seconds; the others are naturally smaller, and each is uncertain by a second or so in any individual earthquake. On the most natural interpretation of the data they indicate thicknesses of about 10 km. for each of the granitic layer and the first two intermediate ones, with uncertainties of about 3 km.

Gutenberg finds similar phenomena in Californian earthquakes; the velocities of the waves P_g, S_g, P_n, S_n are nearly the same as in Europe, but the intermediate ones differ somewhat: he gets 6·05 and 6·83 km./sec. for the P waves. His estimated depths are also near 10 km.

All these earthquakes appear to have had their origins in the granitic layer. This layer is usually covered by a layer of sediments, which are visible at the surface. They have been investigated seismologically to a considerable extent in geophysical prospecting, where the methods of the study of near earthquakes are applied to the waves sent out from an artificial earthquake produced by an explosion. The methods are found to give good determinations of the thicknesses of geological formations, which are verified by actual boring. But in natural earthquakes the sedimentary layer gives little result on account of its small thickness, probably about 2 km. on an average. A P wave in it has been detected in several earthquakes, but is not traceable to great distances. This is probably due to the irregular structure of the layer; every interface between different materials must produce some scattering of a wave in it.

The geological evidence suggests that near earthquakes occurring under the oceans should differ considerably from those in the continents. We should not expect any granitic layer, and the intermediate ones may differ in thickness and even in composition. Hitherto no work has been done on these lines, for geographical reasons. To determine the latitude and longitude and time of occurrence of a near earthquake we need three observing stations recording the same wave; and these must be within 800 km. at the very outside. In the whole of the Pacific Ocean, covering nearly half the earth's surface, the only stations are Honolulu, Apia (Samoa), Suva (Fiji), and Guam. Our chance of being able to carry out a discussion of a Pacific earthquake similar to those that have been done in Europe and California is therefore very slight. Surface waves appear to afford the most promising line of investigation at present. They are certainly less dispersed when following oceanic than continental tracks, which suggests a smaller variation of the velocity of S waves within the upper layers, but an estimate of the thickness must await more knowledge of the velocities.

Surface waves have given additional information about continental structure. As we have remarked already, different periods in these waves should travel with different velocities, and the variation should determine the thickness of the upper layer if there is only one. Stoneley has obtained an estimate allowing for two surface layers, and finds that the dispersion agrees with a thickness of 12 km. for the granitic layer and 24 km. for the intermediate layer.* This is in good agreement with the results from near

* *M.N.R.A.S. Geophys. Suppl.* **2**, 429, 1931.

earthquakes. Further, the near earthquakes studied are all in Western and Central Europe, so that the thicknesses inferred from them refer also to the same region. But Stoneley's data are derived from European observations of surface waves from shocks in Eastern Asia, and therefore are average values for the whole of Europe and Asia. It is important that it should have been verified that the near earthquake results are general and not a purely local condition.

The majority of earthquakes give practically the same times of travel of the P and S waves, within a second or two, irrespective of the position of the epicentre.* This fact shows that the velocities within the earth vary very little so long as the depth is the same. But it is found that the waves in some earthquakes arrive anomalously early at great distances; if the time of occurrence is found from stations 20° away, the P waves at 90° and those through the core may be early by 30 seconds, and the S waves by 50 seconds. This phenomenon was first detected by Professor H. H. Turner; at first it seemed possible that it might be due only to errors in the times of transmission that he was using, these being known to be inaccurate, but with the more accurate times now available the fact is thoroughly established. Turner attributed it to depth of focus. The waves arriving at stations near the epicentre have travelled nearly horizontally, and their times of arrival are affected

* The focus of an earthquake is the place where it actually happens; the epicentre is the point on the surface vertically above the focus.

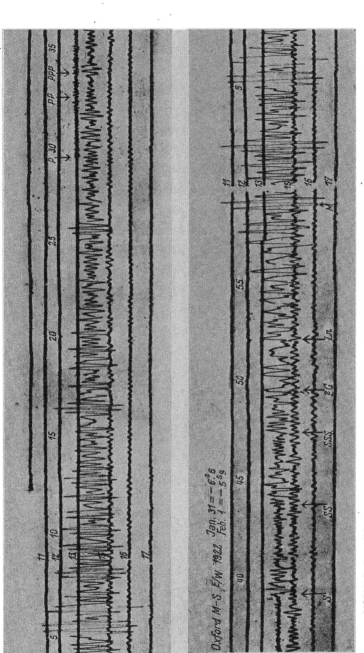

NORMAL EARTHQUAKE, 1922 JANUARY 31.D.13 H. NOTE THE LONG TRAIN OF LARGE WAVES MARKED M.

EARTHQUAKES 49

little if the origin of the waves is at some depth. But the waves arriving at great distances have begun by travelling nearly vertically downwards. A deep focus, therefore, implies that these waves have not so far to travel and consequently arrive after a shorter time. Depths of up to 0·06 of the earth's radius, or 380 km., are fairly frequent. Thus in 1929 the staff of Oxford Observatory, in their work on the International Seismological Summary, studied 592 earthquakes in all: of these, 29 had focal depths of 0·01 of the earth's radius or more. Depths as great as 0·09 of the radius, or 570 km., have occasionally been recognized.

Three independent verifications of the focal depths have been given. The first, due to K. Wadati, depended on the times of the pulses at near stations in Japanese earthquakes. The second, due to Stoneley,* rests on a fundamental dynamical theorem on small oscillations. If a system is disturbed it can vibrate in a number of ways, called normal modes, each with its own period. In any one of these modes there will be a number of points of the system that do not move during the vibration. These are called nodes. If the system is struck at a single point, in general all the normal modes are excited, and any particle of the system moves with the motion due to all the modes together. But if the disturbance is at a node of any mode, that mode is absent from the resulting vibration. The principle is used in stringed

* *Gerlands Beiträge zur Geophysik*, **29**, 417-435, 1931.
4

instruments. One of the harmonics of a string has a period $\frac{1}{7}$ of that of the fundamental, and is discordant with those with periods $\frac{1}{6}$ and $\frac{1}{8}$ of the latter. But by striking or bowing at a node of this mode we avoid exciting it, and there is no discord. Now the surface waves in earthquakes are really an aggregate of normal modes, all of which produce no disturbance at great depths. Hence by this principle these modes should not be excited by a shock at a great depth, and surface waves should be small or unobservable in deep focus earthquakes. This prediction was tested by Stoneley, and found to be correct. The bodily waves are as clear as in normal earthquakes, or more so, but the long train of large surface waves is absent.

The other principle is due to F. J. Scrase.* The waves sent out from the focus of any earthquake undergo reflexion at the surface, and the resulting reflected waves can be recognized. But in a deep focus shock there should be some additional waves arising from reflexion near the epicentre. Scrase finds that these waves are present, and his results have been confirmed by Stechshulte.†

Accordingly the reality of deep foci is thoroughly established.

The records at a single station are often enough to tell us a great deal about the position and time of the earthquake, and even, since the work of Stoneley

* *Proc. Roy. Soc.*, **132A**, 213-235, 1931; *Phil. Trans.*, **231A**, 207-234, 1933.
† *Bull. Seism. Soc. Amer.*, **22**, 81-137, 1932.

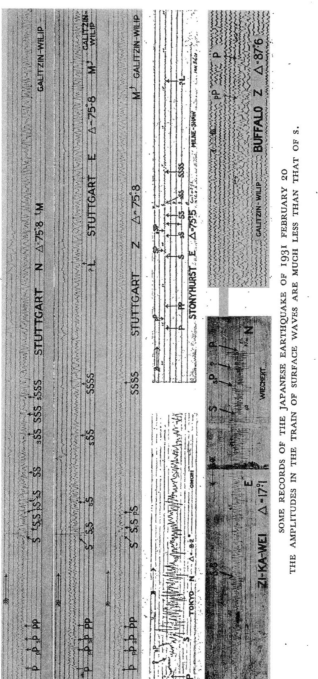

SOME RECORDS OF THE JAPANESE EARTHQUAKE OF 1931 FEBRUARY 20
THE AMPLITUDES IN THE TRAIN OF SURFACE WAVES ARE MUCH LESS THAN THAT OF S.

EARTHQUAKES

and Scrase, about the depth of focus. The interval between the times of arrival of P and S determines the distance from the station. The P movement is always either directly towards or directly away from the epicentre; if both horizontal components are available their ratio gives this direction, so that we know that the epicentre must be at a certain distance away in one of two opposite directions. Usually it turns out that one of these is a region where earthquakes are frequent, the other one where they are rare. Then the more frequent one is chosen; but as soon as possible a number of stations exchange reports by telegraph, and the position is definitely established. Occasional errors arise in such cases as the Baffin Bay earthquake. This region had not produced earthquakes previously, and the Eastern Mediterranean, which is strongly seismic, is at the same distance from Kew in the opposite direction. A single station could only give the Eastern Mediterranean as the most probable position; but fresh information from North America quickly showed that the Baffin Bay situation was the correct one. Cases of deep focus can be recognized at once by the smallness of the surface waves; then the waves reflected near the epicentre are found, and the intervals between the direct and reflected waves give the focal depth.

The cause of an earthquake is in most cases, probably in all, a fracture. All mountains and variations of depth of the ocean floor must have needed distortional stress to produce them, and all need such stress to prevent them from flattening out as water does when it is left to itself. If the distortional stress reaches the strength of the material some type of yield must take place. If this is continuous flow no sudden wave will be sent out. But if fracture occurs the tangential stress across the plane of the fracture is instantly removed, and it was just this stress that was preventing sliding along this plane. Thus we have a sudden change of stress leading to the generation of an elastic wave. We may safely say that every fracture visible at the surface has at some time or other been the seat of an earthquake, the majority of many.

Since the stress leading up to an earthquake is distortional, we may expect that the wave sent out from the focus will be mainly or entirely distortional. It actually seems that in some near earthquakes the primitive P movement is too small to be detected; the observed P movement has been generated from S movement by reflexion at the outer surface. This however is certainly not general; deep focus earthquakes in particular have primitive P movement.

Earthquakes are not uniformly distributed over the earth. They tend to be concentrated along definite belts. One runs along the Pacific Coast of America from Alaska to the Andes, and in the other direction through the island arcs to Japan and the Philippines. Another extends from the Alps through the Balkans and the East Mediterranean to the Himalayas and the Malay Peninsula; the Italian earthquakes occur along a branch from this belt. Off these great belts earthquakes are much less frequent, though they occur in all regions occasionally. But we can be sure that when the mountain systems of Scotland and Wales were being uplifted these countries were much more exciting seismologically than Japan is now. The most active regions at the present time are those of the Tertiary mountain formation, and most present earthquakes can be regarded as the aftermath of that upheaval.

Many attempts have been made to discover regularities in the time of occurrence of earthquakes, but the results are unsatisfactory. For instance, it is sometimes said that earthquakes are more frequent by night than by day. If we take a long period and classify the earthquakes according as they occurred between 6 p.m. and 6 a.m. or between 6 a.m. and 6 p.m., it would be a remarkable accident if the two classes were exactly equally numerous : a mere difference found statistically is not necessarily significant of any general rule. But the difference in a purely random set of times would not be expected to exceed a certain amount; if the difference actually found exceeds this substantially we have adequate grounds for saying that it is real. The same applies to all types of periodicity. Criteria for reality have been given by Sir Arthur Schuster and Sir G. T. Walker. Unfortunately they are seldom applied in published work ; Professor Conrad has applied them to a large number of periodicities that have been obtained at various times, but not one has been found to satisfy the conditions consistently.* It is to be wished that editors of journals would make it an absolute rule not to publish papers on periodicities if these criteria are not applied; if the results are not significant they are worthless, and if they are significant opinion is prejudiced against them in advance. In some cases periodicities appear fairly clearly in observations of the felt movement in earthquakes, but when the instrumental records are examined the periodicity disappears. Except in one respect the distribution of earthquakes in time seems to be wholly erratic and unconnected with any other phenomenon. This exception concerns the fore-shocks and after-shocks of a large earthquake. Such an earthquake is usually preceded by a number of smaller shocks and followed by another series, all from the same place. The number per day increases up to a maximum just before the main shock, and again falls off fairly regularly after it. The fore-shocks thus often provide a warning that a great earthquake is to be expected, though they give little information about the time when it is to be expected. It has also been noticed by Imamura in Japan that earthquakes are frequently preceded by a tilting of the ground, which again serves to predict the place but not the time. This phenomenon recalls Phillips's observation that a stress may not produce immediate fracture, but produces elastic after working which leads to fracture if it proceeds far enough.

*Handbuch der Geophysik, Bd. 4, Lief. 4, 1932.

EARTHQUAKES

The tilting noticed by Imamura may well be this preliminary creep. The after-shocks also suggest elastic after working. The great change of stress occurring in a fracture must lead to a redistribution of stress in the neighbourhood, which may give further creep, leading to other and smaller fractures when it has gone far enough, and these to further ones, until the stress-differences everywhere have been reduced below the limit needed to produce fracture even when left on for a long time.

The actual focus of an earthquake is probably always a small region, not more than a few kilometres in linear dimensions at the outside. If it was larger it would be impossible to separate the later phases on the seismograms as clearly as we usually can. It is sometimes thought that the P and S waves sent out originally can shatter the rocks elsewhere, but this seems to me to be impossible. The stress-difference at the focus was just enough to produce fracture; if it was more the fracture would have occurred sooner. When the earthquake occurs the change in stress-difference spreads out in the elastic waves, but its amount necessarily decreases with distance, and consequently cannot be enough to produce fracture anywhere else, except possibly in regions themselves already on the verge of fracture. The statical change seems to be more important. When a crack is once formed the stresses needed to maintain equilibrium are considerably increased; even in the case of a spherical hole the stresses near it are about twice those at a distance large compared with the diameter, and for elongated cracks the increase is greater. When the elastic waves have been sent out the stress-differences near the focus will therefore in general exceed the strength of the material, even if they were below it before, and the fracture will extend until it reaches regions where the stress-difference is very considerably below the strength. The rate of extension is probably very much less than that of the travel of a distortional wave.

The most conspicuous phenomenon of earthquakes is a movement that can be felt, and in extreme cases may damage buildings. Scales of intensity have been constructed, the degrees of the scale corresponding to the visible or palpable movements. The intensity diminishes with distance, and contours of equal intensity can be drawn. The centre of the innermost contour is near the epicentre. By these macroseismic methods the epicentre can often be fixed within 20 km. or less, so that they give valuable information supplementary to the instrumental data. In most cases the important factor in determining the felt intensity seems to be the horizontal acceleration. It is clear that so long as the ground is at rest or in uniform motion horizontally, a man or a building can stand up vertically without any trouble; it is only when the velocity is varying that any difficulty arises. But when the velocity changes the standing body behaves more or less like a pendulum (the wrong way up), and if the area of support is so displaced that the centre of gravity is no longer over it the body will fall over. A uniform acceleration of the ground can be shown to be equivalent in its dynamical effects to a steady tilt. Thus a body dropped from a height when the ground is accelerated to our right will fall under gravity with its original horizontal velocity, but meanwhile the ground is gaining velocity and the stone will strike it to the left of the point that was originally below the place of projection. Stones dropped at different heights from a vertical building will fall at distances from the foot proportional to the original heights, exactly as if the ground was still and the building inclined to the vertical. Now the construction of a building depends on the principle that the weight of the upper parts is transmitted vertically through the walls, and this is why we make the walls vertical. If the building is tilted, or if the ground is accelerated horizontally, the forces needed to hold it together no longer act through the walls, and stresses are set up in the direction of the floors which may be enough to break the walls. An acceleration of a tenth of gravity, such as often occurs in strong earthquakes, would be equivalent to a tilt of 6°; a man would certainly fall over if he tried to stand at such an angle to the vertical, and few buildings would stand it.

For smaller accelerations the results are less obvious. We may notice at once that the acceleration just mentioned would not necessarily make a man fall over if it was reversed before he had time to fall;

on the other hand, if it was steady he could compensate it by standing at a suitable inclination. To throw a man over the acceleration must persist for a sufficient time, and there is a suggestion that what matters is not the maximum acceleration but the range of variation of velocity. The effect on a building depends on its natural period of vibration, which is recognized when the building sways in the wind. If the period of the ground motion is near the natural period, the motion of the top of the building may be much greater than that of the ground, and the direct effects of acceleration are much magnified. In seismic countries buildings must therefore be designed so that the free periods avoid those that usually occur in near earthquakes, mostly near 1 or 2 seconds. The period can be made short if the building is stiff, long if it is flexible. It must not be made too long, otherwise there is a danger that the building may become unstable under its own weight. In a short-period building what matters is the maximum acceleration; in a long-period one the maximum displacement.

The periods arising in the ground motion are fairly uniform, and probably represent the intervals between various waves reflected internally. Thus the macroseismic observations, however obtained, give some information about the variation of the amplitude of the motion from place to place.*

The actual size of the vibrations varies greatly with distance and with the particular earthquake. In

* Cf. K. Suyehiro, *Proc. Amer. Soc. Civil Engineers*, **58**, 9-110, 1932.

EARTHQUAKES

the Hereford earthquake of 1926 the amplitude in the S_g phase, the largest, ranged from 12 at West Bromwich, distant 63 km., to 0·4 at Zürich, distant 960 km., the unit being a thousandth of a millimetre. This was a small earthquake. In strong Japanese earthquakes amplitudes up to 17 cm. have been recorded. Amplitudes down to a hundredth of a millimetre or less can be felt, though they may escape notice unless conditions are unusually calm in other respects. Damage may arise, however, not only from the vibration but from an effect progressing from one oscillation to another. The San Francisco earthquake of 1906 gave permanent horizontal displacements up to 8 metres of the ground on one side of a fault with respect to the other, over a length of some hundreds of kilometres. This could not be the result of a single oscillation, but of a series of fractures all tending in the same direction and occurring all along the plane of the fault. If a building happens to be across such a fault it cannot escape being torn in two, though a suitably designed one not crossing the fault may survive the vibration. Again, where loose sediments occur, especially on a slope, vibration may encourage settling under gravity, and buildings may collapse under the resulting tilts. Consequently such areas of alluvium often constitute isolated areas of destruction when there is little damage on the firmer rocks around them.

Depth of focus has some relation to the distribution of the intensity, though it has not yet been brought into relation with our knowledge of the structure of the upper layers. We should expect that in a homogeneous crust the maximum displacement in the P and S phases would be nearly inversely proportional to the distance from the focus. If we find a place where the displacement is half what it is at the epicentre, its distance from the epicentre should be proportional to the depth of focus. In comparison with a shallow shock of the same intensity, a deep focus one should be less violent at small distances, but should be felt over a greater area. This is true; for instance the shock of 1926 June 26, which had an epicentre near Cos (Sporades), did damage in Crete at a distance of 250 km. and was felt in Palestine at nearly 1000 km. Its depth of focus is somewhat uncertain, but probably about 100 km. The chief earthquake studied by Scrase was felt up to 1300 km. away, whereas the great Tokyo one was not felt beyond 800 km. The former had a focal depth of 370 km., while the latter was shallow, though a much larger shock. When more material has been analysed it should become possible to determine the relation between depth of focus and the distribution of felt intensity more accurately, and then the macroseismic data should be of great use in finding focal depths. With foci in the upper layers the chief felt waves are probably the train that follows S_g, which spreads horizontally in the sedimentary and granitic layers, and the distribution of intensity is probably quite different.

CHAPTER III
GRAVITY AND THE SHAPE OF THE EARTH

"How is it possible to suspend topaz in one cup of the balance and weigh it against amethyst in the other; or who in a single language can compare the tranquillizing grace of a maiden with the invigorating pleasure of witnessing a well-contested rat-fight?"
—*Kai Lung's Golden Hours*, 258.

WE all learn at the age of eight or thereabouts that the earth is a sphere, as a result of arguments that mostly will not bear much inspection later. Then we learn that it is a spheroid shorter along the polar axis than across the equator. But when we come to consider how the earth's size and shape are measured we find that we do not quite know what these terms mean. Do we, for instance, mean the shape of the ocean surface or that of the solid surface? In what sense can an observer at the top of the Matterhorn say that the earth is a spheroid? A little thought shows that we cannot mean the solid surface; the difference in level between Mount Everest and the deepest part of the Pacific Ocean is about 20 km., and the difference between the polar and equatorial radii is about the same. But the former difference is considered a mere matter of local inequalities, the latter as an expression of a general fact. Yet all our surveys are done on the land; very few observations are made at sea-level, and no measurements of distance at all on the sea itself. On the face of it our surveys seem to be made where the sea is absent, and to describe the shape of a buried surface that we choose to call " sea-level ", but which nobody has ever seen. Clearly the definition of sea-level in land areas must play an important part in our notion of the earth's shape.

The practical definition of sea-level is intimately connected with gravity. A liquid in a tank settles until its surface is flat; we hang up a plumb-line and find that it is exactly perpendicular to that surface. A level surface is one such that gravity at all points of it is perpendicular to it. In this sense the surface of any liquid at rest is level. The surveyor's standard of level is the spirit level: the elevation of a distant object is the inclination of the line of sight to it to the surface determined by the positions of the level as it is turned round. The sea surface itself is not level at any one moment, being disturbed by waves and tides, but these can be dealt with by taking the average height over a long period. This average position of the sea surface at any place is called " mean sea-level ", but even this is disturbed a little by steady forces arising from the action of prevailing winds over the surface and from differences of temperature and salinity. Thus even mean sea-level is not strictly level; the spirit level is better isolated from disturbances, and in comparison with it the mean sea surface shows small but measurable departures from levelness.

GRAVITY AND SHAPE OF THE EARTH 63

When we go from one place to another we find that the direction of gravity changes. In Cambridge I see the Pole Star at an average altitude of about $52°$; but an observer in Spitzbergen sees it at $80°$, one at the equator sees it practically on the horizon, and one in the southern hemisphere never sees it at all, but instead sees groups of stars to the south that are permanently visible to him and never seen from the northern hemisphere. Now these differences are not due to any differences in the actual directions of the stars as seen from the different places; if they were, different stars would certainly be affected unequally, but the angles between the directions of different stars are strictly independent of the position of the observer. The stars are so distant that their directions as seen by all observers are the same. The variation of their altitudes, as judged by the plumb-line or the level, can mean only that the direction of the plumb-line is different at different places. In Spitzbergen the plumb-line is inclined at $10°$ to the direction of the Pole Star; in Cambridge the angle is $38°$, at the equator $90°$, and in Australia greater than $90°$.

Since the introduction of the chronometer, and especially since wireless communication has become general, we can extend this argument to cover differences of longitude as well as of latitude. An astronomer or a navigator (and all navigators are astronomers) knows the Greenwich time at any moment and what stars are then visible at Greenwich. But he himself sees a different set of stars, and the difference again can be attributed only to a difference of the direction of his horizon. An observer in New York sees a star come to the meridian 5 hours later, one in Japan 9 hours earlier, than it does at Greenwich. In New Zealand no star is visible that is visible at the same instant in Spain; similar relations hold between Nanking and Buenos Aires, and between Perth (Western Australia) and the Bermudas. The directions of the vertical at these places are therefore actually opposite. It follows that the earth is at any rate an isolated body with a surface that completely surrounds it. The visible surface does not of course coincide with a level surface, but to a first approximation it does; even on the Matterhorn an inspection of the horizon leads to a general distinction between upwards and downwards which is in fair agreement with that indicated by gravity.

To determine the earth's shape quantitatively we must have recourse to measurement. The standard chosen is the length of a degree of latitude. In the course of a day all stars appear to revolve about a point in the sky, which we call the pole, and which represents the direction of the earth's axis of rotation. The angular elevation of this point above the level surface is the latitude of the place of observation. As we go north or south the latitude changes, and we can measure the distance we have to go to make the latitude change by 1°. If the level surface coincided with the earth's surface the degree of latitude would

GRAVITY AND SHAPE OF THE EARTH 65

be the distance such that the directions of gravity at its two ends differ by 1°. It is found that this distance is on an average about 111 km., but it varies appreciably with latitude. The degree of latitude is shorter near the equator than near the pole by about 1 part in 100. This variation differs somewhat in different longitudes, but it is always in the same direction, and the variation in any one longitude is very much greater than the difference between its amounts in any two different longitudes. The only interpretation of the data is that the earth is approximately a spheroid more strongly curved near the equator than in high latitudes, that is to say flattened along the polar axis.

This ellipticity is an effect of the earth's rotation, partly direct and partly indirect. If the earth was spherical and at rest, gravity at any point outside it would act towards the centre, and the level surfaces would be spheres. But in an ocean rotating with the earth every particle is describing a circle about the axis of rotation, and requires a force at right angles to the axis to make it do so. If the earth is thought of for a moment as a sphere covered by water, and we consider a small portion of water reaching to the surface, this is acted upon by two forces; gravity towards the centre, and the pressure of the other water along the normal to the surface. If the surface was a sphere these could give no component along the tangent towards the poles, so that there would be nothing to keep the water revolving about the axis. But when the surface is an oblate spheroid the two forces are not quite in the same direction, and their resultant gives the requisite acceleration. Actually the body of the earth is also oblate, and this fact is found to intensify the oblateness of the ocean.

We must now consider the effect of height: what do we mean when we say that the height of Mount Everest is 29,002 feet? To understand this statement we must consider the actual method of measuring height. The geodesist making an observation first levels his theodolite by means of a spirit level, and then finds how much his telescope must be tilted up to make a distant mark appear on the cross-wires. The distance of the mark is found by triangulation, and then he calculates the product of the distance and the sine of the inclination. This is interpreted as the difference of height between the distant mark and the observing station. He then proceeds to the mark and makes observations on a higher point, obtaining a new difference of height. The height of the top of a mountain is the sum of the differences of height found as we proceed from sea-level to a place where we can see the top. These differences are measured normally to the level surfaces through the various points of observation, that is, in the direction of gravity. The question of the shape of the level surfaces does not enter into the method.

The height obtained in this way is not strictly unique: we should expect slight differences according to the place we start from and the route we follow.

GRAVITY AND SHAPE OF THE EARTH 67

There is a quantity called the geopotential, however, that is unique. It is the sum of the gravitational potential at a place * and the kinetic energy per unit mass of a particle there due to its rotation with the earth. The difference between the geopotential at two places is the work we have to do on a particle of unit mass to transfer it from one to the other. All points on the same level surface have the same geopotential. If the difference of height between any two successive points of our survey is dh, and gravity is g halfway between the points, the difference of geopotential is gdh. If we add up all the increments of gdh along our survey we get the difference of geopotential between the top of the mountain and sea-level, and this will be the same whatever our route. The sum of the bits dh, however, will not always be the same, because on different routes g may not vary in the same way with h. The practical difference, however, is unimportant, and could easily be evaluated if gravity is known along the route.

Geodesists are in the habit of introducing at this point a surface called the *geoid*, and heights as measured are considered to be heights above this surface. The geoid is defined as coinciding with sea-level on the sea. Where there is land it passes underground in such a way that the geopotential at all points of it is equal to the value at sea-level.

* The attractive force between two masses m, m' is fmm'/r^2, where r is the distance between them and f an absolute constant. The gravitational potential at the position of m' is the sum of the values of fm/r for *all* the surrounding masses m.

If narrow trenches were cut through the land so that the sea could flow in and occupy them all, the surface of the water in these would lie along the geoid. In practice of course this is not done, and the use of the geoid merely introduces unnecessary complications in the theory. For our purposes it is enough, and in theory actually better, to regard the geoid as an inaccessible surface at a depth below the visible surface equal to the measured height.*

The importance of this practical geoid rests on a theorem due to Stokes. Gravity can be measured to a few parts in a million by means of a pendulum. If it is known at all points of a closed nearly spherical surface, and we also know the geopotential at all points, we can determine the earth's gravitational field at all external points. Stokes showed that the problem can be reduced to a simpler one. The observed value of gravity at every place is multiplied by $1 + 2h/a$, where a is the mean radius of the earth; this modification is called the "free-air reduction". We then imagine a level surface with no matter outside it such that gravity at all points of it is equal to actual gravity so modified. There is a definite rule for finding the form of such a surface when the modified value of gravity is known, and then the earth's field is known at all points outside the visible surface. We cannot at once infer anything about the field at internal points, because this depends on the internal distribution of density. The level surface is naturally

* Jeffreys, *Gerlands Beiträge*, **31**, 378-386, 1931; **32**, 206-211, 1932.

GRAVITY AND SHAPE OF THE EARTH 69

taken to correspond to the ocean surface, and it passes at depth h below the solid surface. Some minor corrections are needed for complete accuracy, but Stokes's theory is correct for all quantities depending on the first power of the departure of the form from a sphere, and could be adapted to allow for higher powers if the theory reaches such development as to make this worth while. The chief reason why the theory has not yet been applied in detail is that full observational data about the distribution of gravity have been lacking. But most countries now include the determination of gravity in their surveys, and Vening Meinesz has devised a technique for measuring it in a submarine, which has already been applied over much of the ocean surface.

Knowledge of the field outside the earth cannot by itself determine the distribution of density completely, as we can easily see. A uniform sphere has exactly the same field outside it as either a particle of the same mass at its centre or a uniform shell of the same mass over its boundary. With any external field we can find an infinite number of distributions of density that would give it. But a large number of these are excluded by physical considerations. The density cannot be negative anywhere, and we shall not be ready to suppose that it decreases with depth. When we have excluded distributions that violate these conditions we find that there is a strong family resemblance between the remainder.

Suppose that we have a planet, most of whose surface is a sphere, but that extra mass is piled on a limited region to give a plateau. What is the effect on gravity? Clearly above the plateau the extra mass gives an extra attraction downwards, and at the sides the attraction is towards the mass. A plumb-line hung at the side of the mass is deflected towards it, and the level surfaces, which are at right angles to the plumb-line, slope upwards towards the mass. A level surface that coincides with the surface of the sphere a long way from the plateau will lie above it within the plateau, and the measured height will be somewhat less than the actual height above the sphere. The difference is small for irregularities of small horizontal extent, but considerable for widespread inequalities; small, for instance, for mountains, but important for continents. At the same time gravity above the extra mass will exceed that in a place in the free air at the side, but at the same distance from the planet's centre, by the attraction due to the mass itself. It would not, of course, be fair to compare gravity above the mass directly with that at ground level outside it, for above the mass we are further from the centre of the body and therefore the attraction due to the planet as a whole is less in accordance with the inverse square law. The local variations in the form of the level surfaces are also measurable when latitude and longitude are determined, the spirit level as usual being our standard of uniform height; they are shown by irregularities in the distance corresponding to $1°$ increase of observed latitude or longitude.

On the actual earth such disturbances are superposed on the general field, which includes the effect of rotation and oblateness. We have therefore two types of observational data concerning gravity, one with respect to observed gravity itself, the other to the form of the level surfaces. Their quantitative analysis is complicated. Strictly, if either set of data was complete over the whole surface, it would suffice to determine the other, but as neither is in fact complete they must be used to supplement each other. A general working rule can be given fairly easily. If the extra mass per unit area is $2 \cdot 5 \times 10^5$ grams per square centimetre, it increases gravity above it by $0 \cdot 105$ cm./sec.^{2}. This mass is about that of a kilometre of surface rock. As a rough rule, then, an extra thickness of a kilometre corresponds to an extra $0 \cdot 1$ cm./sec.^{2} in gravity, or 100 milligals in the usual unit. The pressure due to such a load is $2 \cdot 5 \times 10^8$ dynes per square centimetre.

If an excess load, instead of being at the surface, is at some depth, its disturbing effect on gravity is more spread out horizontally.

A remarkable fact of observation, first noticed in the Andes by Bouguer, and since extended to most of the major mountain systems, is that they produce much less disturbance of gravity than would be expected if they were simple added loads. This applies to the Rocky and Appalachian mountains, the Pyrenees, the Alps, and the Himalayas: in fact to all that have been studied from this point of view.

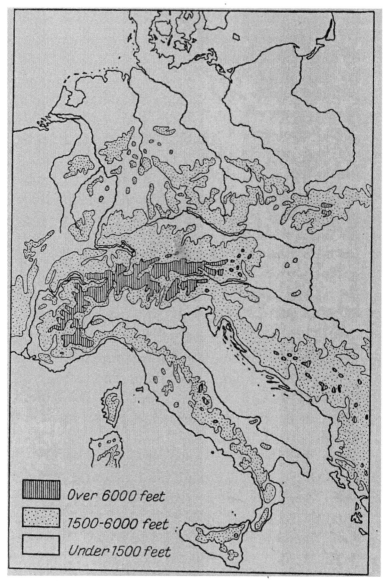

Fig. 3.—Distribution of elevation in Central Europe.

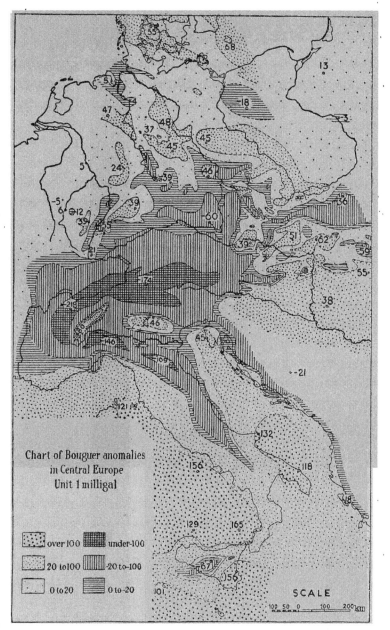

FIG. 4.—Distribution of Bouguer gravity anomalies in Central Europe, showing effect of underground anomalies in density. Compare with heights as shown in map opposite. (*After F. Kossmat.*)

Thus in Central Europe the region where observed gravity is more than 100 milligals less than gravity, calculated on this simple hypothesis, practically coincides with the main mass of the Alps; while spurs of smaller defect of gravity correspond to the Apennines, Carpathians, and the mountains of Bavaria. Meinesz finds over most of the Northern Pacific anomalies of gravity of 25 milligals at most, in comparison with a formula chosen to fit North America; the depth of the ocean reaches 4 km. within a few hundred kilometres of the Californian coast, so that if the Pacific meant simply a replacement of rock by water anomalies of about -300 milligals would be expected. Over the North Atlantic the anomalies of gravity are systematically positive. Now there is nothing in geophysics more reliable than the law of gravitation; the mountains are solid structures and must make their contribution to gravity, and there is a defect of mass over the oceans which must produce a diminution of gravity. If we allow for these effects, which must exist, we are left with the effect of underground matter, which is what we want to know. The result is then that the invisible matter produces an exceptionally low attraction below mountains, an exceptionally high one below oceans. This can mean only that elevations of the surface are associated with reduced density below sea-level, depressions with increased density. It is found in most cases that on the whole the lack of mass below mountains, and the excess below oceans, are about enough to balance

GRAVITY AND SHAPE OF THE EARTH 75

the excess and defect of mass due to the visible inequalities. We call this relation *isostasy*, and the subterranean variations of mass the *compensation*.

The distribution of this compensation with depth is one of the main problems. Hayford and Bowie, in the United States, have worked with the hypothesis that it is uniformly distributed down to a certain depth, called the depth of compensation; below a mountain the density is below the average for the same depth, the defect being the same at all depths. Heiskanen, on the other hand, interprets the phenomena on the lines indicated by seismology and geology. He supposes that there is an upper layer of light material resting on a denser one, corresponding to the upper and lower layers of seismology. Below a mountain range the surface of separation between these is lower than elsewhere, so that a part of the depth that would in plains be occupied by the denser matter is actually occupied by the lighter. Both the American geodesists

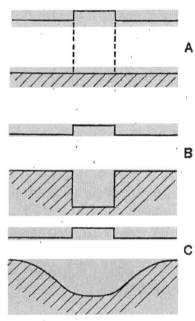

FIG. 5.—Compensation of Hayford, Heiskanen, and regional types.

and Heiskanen assume that the defect of mass due to the exceptional density below is equal to the excess due to the elevation of the surface. On the whole Heiskanen's hypothesis gives slightly better agreement with the observed values of gravity; a great difference is not to be expected, and I think the chief reason for preferring his theory is that it is in agreement with seismic evidence. His estimated depth of the interface is about 40 km., but if, as we may expect, the compensation is somewhat spread out in the horizontal direction, this depth would be reduced. It is in any case in reasonable agreement with the results of the study of near earthquakes.

Isostasy is not, however, complete anywhere, and in some places is not even a first approximation to the facts. In the East Indies Meinesz finds long strips where gravity is abnormally small; changes of gravity of 100 to 430 milligals are to be found within distances of the order of 100 km.* The small values lie near the long narrow strips of great depth that are known as deeps. This would naturally suggest that the deeps are uncompensated; an extra depth of 4 km. of rock replaced by water would mean a reduction of about 300 milligals. But actually the smallest values of gravity do not lie over the deepest places, but some way to the side. To explain the facts we need a rather complicated anomaly of density, probably through a depth of some 40 km. In any case, however, the anomalies are inconsistent with isostasy.

* *Ergebnisse der Kosmischen Physik*, **2**, 153-212, 1934.

In India, again, isostasy gives no representation of the facts, except close to the Himalayas. De Graaff Hunter finds that the form of the level surfaces requires in the Ganges Basin a defect of mass corresponding to 2 km. of surface rock, changing to an excess of 1 km. near Nagpur.* These would correspond to anomalies in gravity of -200 and $+100$ milligals. The residuals in the United States, after compensation has been taken into account, are on an average only about 20 milligals.

These differences in different regions are certainly real; a complete explanation of them is not yet available, but they provide evidence about the present stresses needed to hold the crust in position, which is useful even if we cannot always say how they have come into existence.

In computing a gravity anomaly we compare observed gravity with a standard formula corresponding to a spheroidal level surface with an average value of the ellipticity. If we correct this formula for the height above sea-level, simply allowing for the fact that the place of observation is further from the earth's centre, we obtain the "free-air" anomaly; if we also allow for the attraction of the visible excess or defect of matter we get the Bouguer anomaly (so called because Bouguer was surprised to find it). If we also allow, in our theoretical estimate, for the disturbance due to any form of compensation, we are left with an "isostatic" anomaly. It is unfortunate that in published work the last is often the only one

* *M.N.R.A.S., Geophys. Suppl.* 3, 42-51, 1932; E. A. Glennie, *loc. cit.*, 170-176, 1933.

given. To determine the earth's external field we need the free-air anomaly; while all the information gravity can give about the distribution of density is summed up in the Bouguer anomaly. The isostatic anomaly is really of no interest except to test a particular hypothesis; if it turns out to be unreasonably large the hypothesis is disposed of, and the isostatic anomaly is of no use until we have undone the work of calculating it.

The general variation of gravity with latitude at sea-level is intimately connected with the ellipticity, and in fact gives our best means of determining it. But it also gives information about the distribution of density in the interior, which, when combined with the evidence of seismology, provides full knowledge of the elastic properties all the way to the centre. The associated part of the earth's field can be summed up in its moments of inertia.* When the moment of inertia is known, the mass being already known from the mean value of gravity, we have information about the distribution of density along the radius. But we have other information from seismology, namely, that there is one great discontinuity of properties in the earth about halfway to the centre, with which no other change is comparable. As a trial hypothesis we assume that the earth has one density down to this level, and another below it, and we choose the densities to make the mass and moment

* The moment of inertia of a body about any line is found by taking each part of its mass, multiplying by the square of its distance from the line, and adding all the results.

GRAVITY AND SHAPE OF THE EARTH

of inertia right. The result is that the shell has to have a density of 4·3, and the core 12. This, however, is not all, since both core and shell are heavily compressed by the weight of the overlying material; but the velocities of the elastic waves tell us how much the compression would be. If the pressure was taken off, the mean densities would fall to about 3·4 and 8, nearly those of olivine and iron at ordinary pressures. These results are very reasonable; the seismic data, which depended on the velocities of the waves, not on the density, pointed to olivine for most of the shell, and iron is the commonest heavy metal at accessible depths.

Other evidence is provided by the moon. We should expect it to be composed of much the same materials as the earth, though its smaller size may go with a great difference in their proportions. Its density is 3·33, and astronomical data make it difficult to suppose that it has any appreciable central condensation, as De Sitter has shown. But on account of its smaller size and gravitational effect we should expect its matter to be compressed very little; the density at low pressures would be 3·2 or 3·3, which would again agree with olivine as the chief constituent.

Accordingly we have a fairly complete idea of the constitution of the earth. Most of the rocky shell is probably an olivine, while the core is most probably liquid iron. Near the surface there are layers of lighter rock, the uppermost in the continents being sedimentary and granitic. The intermediate layers in the continents and the upper ones below the oceans can be identified at present with much greater doubt.

The ellipticity adopted for the earth, however, represents only the largest part of its deviations from a spherical form. Apart from mountain ranges, which we can regard as local phenomena, there are widespread anomalies in gravity, of small amount, but keeping the same sign over great distances. Thus Meinesz finds positive anomalies averaging about + 30 milligals right across the North Atlantic. There is a systematic variation in longitude, which Heiskanen expresses by an ellipticity of the equator. The standard level surface, in other words, is an ellipsoid with three different axes instead of a figure of revolution. We have not enough data at present to find the true form, gravity not having been measured over the whole of the sea surface in the southern hemisphere. But the existing observations are quite enough to establish the existence of these inequalities, though there may be doubt about their distribution. It is quite impossible that they can be isostatically compensated at the usual depth of 50 km. or anything like it. If they were, the variation of observed height needed to explain them would have to be some tens of kilometres, which is out of the question, since the observed height is not noticeably correlated with the gravity anomalies.*

A similar inequality has been known since the time of Laplace to exist in the figure of the moon. If a body that is not spherical is acted on by the attraction

* *M.N.R.A.S., Geophys. Suppl.* 3, 69.

GRAVITY AND SHAPE OF THE EARTH

of an external body, the total attraction does not act through the centre of mass; there are also couples tending to turn it about the centre of mass, whose magnitudes depend on the differences between the moments of inertia about different axes. It is for this reason that the attraction of the sun and moon on the earth produces the precession of the equinoxes and its associated nutations. In the moon the visible results are different on account of the slower rotation, but they are clearly observable. One effect is that the inclination of the moon's axis to the plane of its orbit is kept constant, in such a way that while the pole of the moon's orbit revolves about that of the ecliptic in 19 years, the moon's axis of rotation keeps in the same plane as these two, and at a constant angle to both. The magnitude of this angle determines the difference between its greatest and least moments of inertia, the greatest being about its axis of rotation and the least about an axis pointing nearly to the earth. This difference is 16 times what it would be if the moon was of uniform density (which is not far wrong) and the outer surface was a level surface. The natural explanation in the light of the theory of tidal friction is that the moon solidified when it was at about 0·4 of its present distance from the earth, and has ever since maintained the figure it had then; the tide raised in it at that stage by the earth has persisted ever since, being maintained by the strength of the moon's material. But the immediate consequence is that the inequality of external figure is one that produces a disturbance of the moon's gravitational field and therefore cannot be isostatically compensated, except at the cost of an enormous increase in the inferred ellipticity of the outside.

On the classical theory of the figure of the earth, which is largely due to Clairaut, the earth was supposed to be almost entirely fluid, with only a thin solid crust. The argument was effectively that the temperature rises by about 30° for each kilometre of depth, and at this rate the melting-point of any reasonable material would be reached at a depth of 50 km. Below that level the matter was supposed to be fluid, and a theory of the ellipticity of the layers of equal density such that it could remain in steady rotation under its own attraction could be constructed. Even with our present knowledge the theory still needs to be taken seriously, because if we work out the consequences of supposing the earth to be devoid of any permanent strength, the differences between them and the observed facts give us a means of finding how far the theory is wrong; that is, we can estimate the earth's strength. On the earth we can observe three quantities associated with its figure: the mean value of gravity, the ellipticity (or the variation of gravity with latitude) and the precessional constant $(C - A)/C$, where A and C are the moments of inertia about the equatorial and polar radii respectively. By a theorem due to Radau it turns out that the distribution of density with distance from the centre makes hardly any difference to the results. If we

GRAVITY AND SHAPE OF THE EARTH 83

choose any distribution that makes the precessional constant right, the ellipticity inferred for the outside will always be practically the same, and so will the mean moment of inertia. As the precessional constant is known to a very high accuracy it gives a very accurate determination of the ellipticity, which is in good agreement with the best geodetic values, though these have a larger uncertainty. But the theory leads to other consequences that are certainly wrong, namely that the other widespread inequalities in gravity are impossible; it allows no variation in surface gravity except the symmetrical elliptic one. Nevertheless it is interesting that, while gravity varies by about 5000 milligals from equator to pole, the theory of the fluid earth represents this variation to within about 6 milligals, which is less than the widespread irregularities; it would have been very disturbing had the other variations of the order of 30 milligals been associated with an error of, say, 1000 milligals in the result given by the theory of the fluid earth for this largest inequality.

The general result is, then, that the supposition that the interior of the earth is under hydrostatic stress alone gives a good approximation to its external figure; but there are real departures which can only be attributed to an inaccuracy in the hypothesis; in other words, the stress is not exactly hydrostatic, and stress-differences exist. The evidence of the theory of the figure of the moon gives us every reason to believe that these can be supported for intervals of time comparable with the whole age of the earth.

CHAPTER IV

THE STRENGTH OF THE EARTH

"'It is well said that after passing a commonplace object a hundred times a day, at nightfall its size and colour are unknown to one', replied Pe-Lung."
—*Kai Lung's Golden Hours*, 196.

THE existence of any differences of height on the earth's surface is decisive evidence that the internal stress is not hydrostatic. If the earth was liquid any elevation would spread out horizontally until it disappeared. The only departure of the surface from a spherical form would be the ellipticity; the outer surface would become a level surface, the ocean would cover it to a uniform depth, and that would be the end of us. The fact that we are here implies that the stress departs appreciably from being hydrostatic; if it was hydrostatic the inequalities of the surface would spread round the earth in the time it takes a gravity wave to travel from one elevation to the next depression, a matter of a few hours at most in a mass the size of the earth. The amount of the departures gives an indication of the stress-differences needed to support them. They imply differences in the load (weight per unit area) over a level surface near sea-level; our problem is to find out what stress-differences these imply in the interior.

The problem has no unique answer. If the stress in the earth was originally hydrostatic and the earth was an elastic solid, we could find out just what stresses would be produced in the interior by given surface loads; this problem was solved by Sir G. H. Darwin for a uniform earth. It has applications to the changes introduced by denudation and deposition. The stress in the interior of a plateau may be approximately hydrostatic to start with, and so may that near the mouth of a river. But when rivers cut deep valleys in the plateau the load is reduced in the plateau by the weight of material carried away, and when it is redeposited it produces a corresponding increase in the load in its delta. The change of load produces elastic deformations, which give new non-hydrostatic stresses in the interior. If the material has a finite strength these will persist until the change of load has been enough to make the stress-differences reach the strength; then fracture or flow will take place. The loaded region is then insufficiently supported and will sink, while the valley will rise. At this stage the pure elastic theory has of course broken down.

But there is no reason to suppose that the theory is applicable to the inequalities represented by mountains in comparison with plains, and continents in comparison with oceans. A mountain system has not been produced by the deposition of extra mass on the surface; it is the result of a complicated movement through a considerable depth. This movement has visibly involved fracture and contortion and therefore must have arisen from the existence of stress-differences. There were stress-differences before the mountains, and the elastic stresses due to the weight of the mountains, while perfectly real, are not superposed on a hydrostatic stress, but on a system in which part of the original stress-differences has almost certainly survived. In other words, if the whole of the Alps were planed down to sea-level, the internal stress-differences would not sink to zero; it is probable that enough new elastic stresses would be produced to make new fractures. We have no means of saying how the present stresses are distributed; if one distribution can be found that would give equilibrium, there will be an infinite number of others. This ambiguity can be understood if we consider a rectangular block exposed to pressure on the ends. It is compressed to a definite shape, with a definite stress-difference inside; but this shape might equally well be the natural shape of another body with no internal stress at all.

We can, however, be quite sure that stress-differences exist, merely because mountains and continents do not disappear in a few hours. Many different arrangements of the internal stress would be capable of supporting them, but they all involve stress-differences. In any one of these distributions there will be a place where the stress-difference has a value not exceeded anywhere else, but with some this maximum will be greater than for others. We can now ask how great this maximum needs to be; if every system of stresses

THE STRENGTH OF THE EARTH 87

capable of supporting the visible inequalities implies that a certain value of the stress-difference is exceeded somewhere, we can assert that the visible inequalities imply that the earth is capable of supporting at least that stress-difference. In other words, we obtain a lower limit to the strength of the earth.

The solution depends not only on the maximum surface load, but on its horizontal distribution. The cases that have been investigated are, first, harmonic loading, which represents a series of parallel ranges and valleys; second, a single range with precipitous sides; and third, a single range sloping gradually from the centre.* In the last case we obtain the surprising result that with a given maximum height the load can be supported with as small a maximum stress-difference as we like, provided the load is distributed sufficiently widely and in a special way. But this is the kind of exception that proves the rule; we can support one range in this way, but not two, unless they are at an impossible distance apart. In all cases with any resemblance to actual ones we get very much the same result: the maximum stress-difference has to be about a third of the range of load. Thus if a mountain stands at a height of 5 km. above the nearest plain or the bottom of the nearest valley, the range of load, assuming a density of 2·5 for the surface rocks, is $980 \times 2\cdot5 \times 5 \times 10^5$ or $1\cdot2 \times 10^9$ dynes per square centimetre, and the strength needed to support it is 4×10^8 dynes/cm.². For the great

* *M.N.R.A.S., Geophys. Suppl.* 3, 30-41, 60-69, 1932.

oceanic deeps the strength needed is greater, both the range of depth and the density being more than we have just supposed. Now these values are within the range indicated by laboratory measures of the strengths of rocks. The stress needed to crush a column of granite is about 10^9 dynes/cm.^{2}, for basalt somewhat more. It appears therefore that if this strength is maintained to a sufficient depth there is no difficulty about accounting for the support of surface inequalities. In all the cases so far considered the maximum stress-difference occurs at a depth about equal to the horizontal distance between the highest and lowest points of the surface. If a mountain ridge is 2 km. from the nearest valley, the maximum stress-difference is at a depth of about 2 km., and so on; at greater depths the stress-differences are smaller.

The phenomena of isostasy indicate that these results need some modification for the larger inequalities of height; we know the surface load, but we also know that there is a deficiency of mass below the greater mountain ranges, and that below a depth of about 50 km. most of the variation of load has disappeared. When the deficiency was noticed in the Himalayas by Pratt, an explanation was at once suggested by Airy. At that time the earth was generally believed, on account of the temperature gradient measured in mines, to be liquid at depths of 50 km. and more. This is now known not to be true, but it remains possible that the material at these depths may be very much weaker than similar material

THE STRENGTH OF THE EARTH 89

would be at the surface. Then we can find the effect of loading the outer surface over a region whose horizontal extent is large compared with the depth of the strong layer, on the supposition that the lower layer has the greater density. The crust will be bent down under the load, so that its lower surface is more deeply immersed in the lower material. If the lower material is too weak to behave elastically, it is displaced outwards like a fluid; but as the pressure increases with depth the depression of the crust brings in an extra pressure over the interface, which balances the surface load when the depression is great enough. But when the lower material is driven out below the load the mass per unit surface is reduced, and it turns out that balance is attained when the loss of mass due to the expulsion of the lower matter nearly balances the extra mass on the surface. This result is true as an approximation whether the crust breaks or merely bends. In either case it explains the observed compensation of mountains; below a surface load there is a nearly equal defect of mass at some depth, due to the replacement of part of the denser matter of the lower layer by an extra thickness of the lighter upper layer. The mountain mass may be likened to an iceberg; the upper surface is elevated, but the conditions of floating require that its base also projects downwards into the sea, the extra mass on top being just balanced by the deficiency below sea-level.

We can explain in this way the smallness of the disturbance of the earth's gravitational field by mountains and continents as due to the weakness of the lower layer, this in turn being due to its high temperature.* The actual stresses in the upper 50 km. remain arbitrary to a large extent; as we have just remarked, the depression of the interface and the effect on gravity will be very nearly the same whether the crust breaks or merely bends, but there is a great difference in the stresses in the crust in the two cases. We must therefore proceed as we did for a uniform earth, and ask the following question. Given the surface inequalities, together this time with the principle that the stress-differences are zero at depths over 50 km., what is the smallest strength at depths less than 50 km. that would suffice to maintain the surface inequalities? Evidently if the surface inequalities are of small horizontal extent the extra condition makes little difference. If the distance from a mountain top to the nearest valley is only 5 km., the load can be supported elastically by stress-differences of about one-third of the range of load to a depth of about 5 km., but at greater depths the stress-differences rapidly become less. It makes little difference if we suppose the stress-differences to be zero below 50 km., because in this case they are negligible at these depths anyhow.

The change, however, is serious for loads of great

* It is here that I differ fundamentally from Wegener and his followers. They consider weakness as characteristic of the lower material and not due to difference of temperature, and then proceed to infer a weakness of the ocean floor that is contrary to all direct indications.

THE STRENGTH OF THE EARTH

horizontal extent. Suppose the centre of a mountain mass is 200 km. from the nearest plain. If we were free to distribute the stress-differences as we liked, we could support such a load with the greatest stress-difference equal to about a third of the range of load as before; but the greatest values would occur at depths down to 200 km. In the present case our extra condition excludes the existence of stress-differences at the depths where the simpler theory needs them most, and equilibrium can be preserved only by means of greater stress-differences than before, which can be only in the top 50 km. It turns out that this can be done with the smallest possible stress-differences in the following way. There must be no tangential stress across vertical planes, that is, one of the principal stresses at any point is vertical. The weight of any surface load is simply transmitted vertically to the interface, where it is supported by the extra pressure in the lower layer. The stress-difference in the upper layer is then equal at all depths to the excess load. This condition determines the depth of the interface to a closer approximation than before, and is found to be practically equivalent to the condition that the mass in a vertical column is independent of the load. If we try to support part of the load by tangential stresses over vertical faces the stress-differences become much larger.

We can now estimate the strength needed to support a compensated mountain system. If we take the average height of the Himalayas as 5 km. the excess load is about $1{\cdot}2 \times 10^9$ dynes/cm.^{2}, and this is the strength needed to support their general form. Within them there are deep valleys, but the average distance between these is only of order 100 km., so that the most conspicuous relief can be supported by stresses at depths less than 50 km. and the extra strength needed to support them is only about a third of the range of load. This may be important, because we do not need as much strength to hold up Mount Everest, the top of which is only 70 km. from the Arun gorge, as we should if the whole Himalayan area was at the level of its highest point. In any case, however, it seems that the Himalayas require the full strength of surface igneous rocks to support them; there is no room for any reduction of strength with depth in the upper layer, and we may have to infer an increase.

It is important to emphasize the fact that the mechanism of compensation implies a greater strength in the upper layers than we should need without compensation; many writers have too readily extended the probable weakness of the lower layer to the upper ones. Compensation means that the lower layers are not contributing by their strength to the support of the visible inequalities, and consequently the upper ones have to do more, not less. If the upper layers were also weak there would be no surface inequalities at all.

The supposition that there are no stress-differences in the lower layer is, however, only approximate.

THE STRENGTH OF THE EARTH 93

It provides an explanation of the general distribution of gravity and of its behaviour near large mountain ranges, but at the best there are still anomalies far exceeding the errors of observation, as we remarked towards the end of the last chapter. The widespread inequalities of gravity can be interpreted only as due to a real variation of the mass in a vertical column, and to support this by stress-differences in the upper layers alone would require an impossible strength. Such a mass may be compared with one resting at the end of a thin beam clamped at the other end; if the beam is supported below, the stress-differences are confined to the neighbourhood of the load and are comparable with the load; but if the beam is free it will be bent much more, much greater stress-differences will be required throughout its length, and it may break near the fixed end. These inequalities of gravity therefore imply a variation of load on the lower layer, and require stress-differences in the lower layer. To make these as small as possible, they must be distributed through a depth of some thousands of kilometres on account of their great horizontal extent, and therefore through the whole of the earth's rocky shell. The strength needed is about 5×10^7 dynes/cm². This is only $\frac{1}{24}$ of what we have inferred for the upper layers, but it is not negligible. The inequalities in India and those found in the East Indies by Meinesz are of smaller horizontal extent but of larger magnitude, and speak for a strength comparable with that of the upper layers through a depth of at any rate some hundreds of kilometres. The excessive ellipticity of the moon implies internal stress-differences of at least 2×10^7 dynes/cm². in its interior.

The significance of these values can perhaps be best visualized by comparing them with the measured crushing strengths of building materials, as given by Unwin.* His data, expressed in the present units, are as follows:—

Material	Strength (10^8 dynes/cm.²)
Basalt	11·6 to 17·0
Granite	7·7 to 17·3
Limestone	4·3 to 15·7
Sandstone	1·86 to 6·2
Bath Stone	1·02
Brick	0·93 to 5·0

The strengths of the stronger basalts, granites, and even limestones approximate to that required to support the greater mountain systems, but without much to spare. The strength inferred for the greater part of the rocky shell is about half that of Bath Stone or the weaker bricks. Material with such a strength would still be serviceable for building purposes. Assuming a density of 3, a column 170 metres high could stand without crushing under its own weight. The corresponding value for the strongest basalt would be 5·6 km.

The existence of an appreciable strength in the lower layer is confirmed to some extent by the existence of deep focus earthquakes. The bodily waves in many of these are as strong as in the greater normal earthquakes, and we may reasonably suppose that the stress-differences at the foci, the release of which gives rise to them, are equally as great. A

* *The Testing of Materials of Construction*, 1888.

THE STRENGTH OF THE EARTH 95

material devoid of strength would never have accumulated such stresses; it would have broken or flowed before they were reached. As it happens, the region of large gravity anomalies found by Meinesz in the East Indies is subject to deep focus earthquakes, and the same may apply to all the Pacific deeps. The strength here at depths to 300 km. or so may be much greater than usual, so that these places may be abnormal. But the earthquake of 1926 June 26, in the East Mediterranean, and that of 1929 February 1, in Afghanistan, had focal depths of about 140 and 160 km. respectively, so that even in mountainous regions and in comparatively shallow seas there is evidence of strength at great depths.

To sum up, the observed distribution of height and of gravity requires strengths in the neighbourhood of those of the strongest surface rocks down to depths of about 50 km. There is evidence that the strength in most regions falls off considerably below that level, but it must remain appreciable through most of the rocky shell. It is possible that in the regions subject to deep focus earthquakes the reduction of strength does not begin until depths of the order of 300 km. are reached.

It is natural to attribute the decline of strength with depth to the rise of temperature, but our experimental data are not sufficient for a quantitative determination, and while a theoretical determination may be possible none has yet been given. We are therefore not in a position to estimate the distribution of temperature from that of strength, or conversely, except in qualitative terms.

CHAPTER V

RADIOACTIVITY AND THE EARTH'S HISTORY

"As to this, says the Wise One, 'When two men cannot agree over the price of an onion, who shall decide what happened in the time of Yu?'"
—*Kai Lung's Golden Hours*, 76.

So far we have been concerned with the earth's present state. Both in the earth itself and in its astronomical relations we find reasons for believing that this state is changing, slowly by human standards, but quite definitely. Rivers are continually carrying sediments and dissolved salts to the sea, derived from the erosion of rocks on the land. Some of these rocks are sedimentary, but igneous rocks are exposed at the surface over large regions, and when an igneous rock is eroded there is no way of reassembling the constituents into a rock similar to the original one. The particles of quartz and much of the felspar and mica are deposited in the sea and in lakes to form sandstone; the clay derived from the decomposed felspars gives shale. The chief dissolved metals are sodium and calcium. Calcium in the sea goes to form the skeletons of living organisms, such as shells, corals, and foraminifera. When these die and become buried under others they form limestone.

RADIOACTIVITY AND EARTH'S HISTORY 97

Sodium, however, remains in solution and is converted into salt. Thus the total quantity of sedimentary rocks on the earth's surface is continually increasing ; so is the salt in the sea. As we look back in time, therefore, we see the oceans fresher and fresher, and the total amount of the sedimentary rocks less and less, and we have an indication of an early time when the whole surface was covered by igneous rocks and the sea was fresh. Estimates of the times needed for these changes, on the supposition that the rates of transfer to the sea have been uniform, are about 300 million years ; but these can be taken only as suggesting an order of magnitude, because there is every reason to believe that the present rate of erosion is abnormally high.

The source of the chlorine that helps to form the salt in the ocean is somewhat of a mystery. Chlorine is not abundant in rivers ; most of the sodium is in the form of carbonates or sulphate. But chlorine is prominent in the free state and as hydrochloric acid in the gases emitted from volcanoes ; and an ocean containing sodium carbonate would certainly absorb these gases from the air until either the whole of the sodium, or the whole of the chlorine, had been converted into salt. What is not clear is why there should be so little excess of either element, seeing that their sources are quite independent. It is not as if hydrochloric acid was the chief agent of chemical denudation ; then we might expect the sodium extracted from rocks to balance the chlorine, but then the balance would hold in rivers as well as in the sea, and it does not. The problem is very serious for biology, for if either sodium or chlorine had ever been strongly in excess all life in the ocean would have been destroyed. It seems that either the excess of sodium or that of chlorine must act directly on the ocean floor and extract enough of the other to maintain the balance, but it is not clear how this happens. A small variation of the supply of either would not affect neutrality, because the ocean contains also a good deal of magnesium and calcium as chlorides. A temporary deficiency of chlorine would precipitate these metals as carbonates, while an excess would redissolve the limestones until neutrality was restored. But an excess of chlorine large enough to dissolve all the limestones, or one of sodium enough to precipitate all the calcium and magnesium, would make the ocean too alkaline or acid, and life in the sea would have to start again from scratch.

Analyses of igneous rocks and sediments show that in the conversion of the one into the other about two-thirds of the sodium is lost, presumably to reappear in the sea; thus the total amount of sodium in the sea gives information about the total quantity of sediments, which is enough to cover the continents to an average depth of about 2 km.* The actual depth certainly exceeds this in places, but this device is the only clue at present to the average thickness of the sedimentary layer.

* Jeffreys, *Gerlands Beiträge*, **26**, 58-60, 1930.

RADIOACTIVITY AND EARTH'S HISTORY

The tides raised in the ocean by the sun and moon are subject to a continual loss of energy due to fluid friction; this energy has to be continually replaced from the earth's rotation and the moon's motion. The effect is that the earth's rotation is becoming slower and the moon going further off. The effect of friction on the tides is observable in shallow seas such as the Irish Sea, the English Channel, and the Yellow Sea, and its amount just explains an apparent steady acceleration of the moon's motion which used to present a serious difficulty in the lunar theory. If we look back in time, we see the moon closer to the earth and the earth rotating more rapidly; but there is a minimum possible distance between their centres, which is about twice the earth's radius. If the distance was ever less than this the moon would have approached the earth and fallen into it. If we take the present rate of loss of energy as a guide and allow for the more rapid rate in the past due to the smaller distance, we find that the moon must have been at its minimum distance about 4000 million years ago. This estimate is only a rough one, because the distribution and area of suitable seas have certainly changed, but it does show that we cannot take the history of the earth back more than a certain interval of time. If the moon was never so near, the possible age is correspondingly shortened.

Other limits to the age of the earth are imposed by the phenomena of radioactivity. The heavy elements uranium and thorium have a habit of breaking up to give lighter elements. The final result of the break-up of an atom of uranium, of atomic weight 238, is an atom of lead of atomic weight 206 and 8 atoms of helium of atomic weight 4. An atom of thorium, of atomic weight 232, gives an atom of lead of atomic weight 208 and 6 atoms of helium of atomic weight 4. The two kinds of lead differ in atomic weight but have the same chemical properties; they have been separated by Aston, who uses a method depending directly on the mass. Both are present in ordinary lead, of average atomic weight 207·1, together with a third kind of weight 207. The rates of the transformations are known, and are found to be independent of temperature and pressure. In a specimen of uranium one atom in 6600 million breaks up every year; for thorium the rate is more difficult to measure and is less accurately known, but seems to be about one in 19,000 million per year. If, therefore, we have a specimen of what was once a pure uranium compound and determine the present amounts of uranium and either lead or helium in it, we can find how far the disintegration of the uranium has proceeded and hence the age of the compound. The same can be done for thorium compounds. Unfortunately, in spite of the obvious advantages of such a method in determining the time since a given piece of mineral crystallized, there are a number of practical difficulties. If lead was present to begin with, the lead we find is not all the result of radioactive processes, and the estimate of age will be too high. If lead has been removed by percolating water, the age will be underestimated. If we use the helium method, some of the helium has probably been lost by diffusion, and more in the actual preparation of the mineral for analysis, so that ages found by the helium method are systematically too low. This method has consequently received less attention than that based on lead. In the lead method we must confine attention to minerals that contain little original lead and have been unaffected by changes due to external causes. If these conditions are not satisfied no useful determination can be made. It is possible to test for original lead by an atomic weight determination or by Aston's method; external changes are perhaps more insidious, especially for thorium compounds. Holmes has pointed out that lead formed in a uranium mineral would form lead uranate, which is very insoluble and would stay where it is; but lead in a thorium mineral would form oxide, which is fairly soluble and would be carried away if water reached it.

In spite of these difficulties a large number of good determinations have been made. The technique is most fully described by A. L. Kovarik and the results by A. Holmes in the recent Bulletin of the U.S. National Research Council on the *Age of the Earth*. Minerals with ages up to 1500 or 1800 million years are known; the Cambrian period, which contains the oldest identifiable fossils, began about 450 million years ago; the Permian period was about 200 million years ago, while the Tertiary era began about 40 million years ago. These ages are obtained from the lead ratios. Many more ages have been determined in this way, but it is often difficult to attach definite positions in the geological time-scale. The lead ratio gives an absolute measure of age, but it requires the existence of igneous rocks containing minerals with a high content of uranium, which are not common. Thus it can be applied to only a fraction of the igneous rocks that exist. Further, to find the geological date of an igneous rock is one of the most difficult problems of geology. Stratigraphical geology depends on the fact that sedimentary rocks are deposited in layers, each upon the preceding one; if three beds are found in order A, B, C, B resting on A and C on B, we can say that A is older than B and B than C, but as a rule we have no means of saying how much older. Approximate equality of age of rocks in different parts of the earth is established by fossils. When a new species of animal or plant comes into existence it spreads rapidly over the surface where conditions are favourable to its survival; when we find the same new type appearing for the first time in two widely separated places we can say that the first rocks containing it in the two places are of the same age. Presumably they cannot be exactly the same age, but the time needed to spread appears to be short compared with a geological period. If it were not, and we considered two types originating at the same time in Europe and North America, we should find the European type at the lower level in

Europe and at the higher level in North America. This does not happen. We are therefore entitled to suppose that spreading is rapid enough to be considered geologically instantaneous. A species may perhaps take millions of years to reach all the places accessible to it and adapted to its survival, but a million years is a short time in geology. But we must notice that to fix the geological date accurately depends on the presence of sedimentary rocks with identifiable fossils. When a region is above the sea for a long period the only organisms found in it will be in occasional lake and river deposits; in a desert even these will be absent. Thick beds of rock without fossils are often formed even in the sea. Thus there are long gaps in the geological record at any place, which can be filled in only by comparison with other places. Now what happens in the case of an igneous rock poured out over the surface? It perhaps lies on one set of fossiliferous rocks, and is therefore newer than these; another set lie on top of it, so that we have upper and lower limits for the geological age of the igneous rock. If the rocks above and below contain the same fossils, the difference of age is negligible. If, further, the igneous rock is obliging enough to contain a uranium mineral, we can determine its absolute age, and therefore we have the absolute age corresponding to a particular geological age. With a sufficient number of such determinations we could calibrate the geological time-scale, and the dates of intermediate points could be filled in with fair accuracy by interpolation on reasonable assumptions about the rate of deposition of sediments. In fact, however, these conditions are seldom satisfied. The igneous rock may be deposited on a land surface and may lie exposed for ages before any fossiliferous sediments are formed on it, so that its geological age may be known only within wide limits.

In the pre-Cambrian era, when no fossils are found, the geological difficulties are even greater. The only way of establishing identity of age between sediments a long distance apart is actually to trace a given formation all the way. This task is necessarily slow and troublesome, especially since these old rocks are usually altered from their original condition. Nevertheless with modern technique much progress can be made; correlation has been achieved, for instance, right across the Highlands of Scotland.* On the other hand, there is no way of connecting Scotland with North America. But if we can find the absolute ages we can establish identity of age even at these distances, so that in the pre-Cambrian era the lead-uranium ratio is becoming a method of pure geological technique.

The helium-uranium ratio has attracted less attention than the lead-uranium one, since it was found that it gave ages that were systematically too low. It appears now, however, that it may be useful for dating igneous rocks containing little uranium, and if so it has a great advantage over the lead method.

* H. H, Read, *Trans. Roy. Soc. Edin.*, 55, 1928.

RADIOACTIVITY AND EARTH'S HISTORY

The latter is useless unless the uranium content is high, since the original lead will be comparable with that formed from uranium. But helium is a gas, more mobile than any other except hydrogen, incapable of forming chemical compounds, and indisposed to enter into solution. For all these reasons original helium is unlikely in a rock; when the rock was fused the helium would rise to the top and escape into the air. Again, in a rock containing a small but generally diffused amount of uranium, it does not matter much if the helium generated has spread out by diffusion. This only makes for uniform concentration, and if the uranium itself is uniformly spread the helium will be uniformly spread when it it formed; there is nothing for diffusion to do. A method based on these principles has been introduced by V. S. Dubey and A. Holmes within the last few years. It has given determinations of the ages of the Whin Sill of the North of England (end of the Carboniferous) and of the Cleveland Dyke (Tertiary), and seems likely to become one of the most valuable practical methods.

These methods have two uses: they assign absolute dates to geological events and hence give a time-scale for geological processes, and they give a lower limit to the age of the earth, which must exceed that of the oldest rocks known. We can say quite definitely that the age of the earth is more than 1500 million years. There are also two methods based on radioactivity that fix an upper limit. We are in the habit of thinking of uranium and thorium as rare elements, but this is really only because they seldom occur in such concentration as to make it profitable to extract them. The average amounts by weight are respectively about 6 and 15 parts in a million. A cubic metre of average rock of density 3 will contain 18 grams of uranium and 45 grams of thorium. But lead also is generally distributed. Some of this lead is presumably original; hence if we proceed on the supposition that the whole of the lead has come from the uranium and thorium during the existence of the earth we shall overestimate the age. This device was introduced by H. N. Russell. With some corrections introduced later by Holmes and myself it is found to give a maximum age of about 3000 million years.

The other line of argument is due to Rutherford. The isotope of lead found by Aston, with atomic weight 207, cannot be derived from either uranium or thorium, which give lead with atomic weights 206 and 208. Yet Aston finds lead of weight 207 even in uranium minerals. Now uranium is found to be associated with a minor series of radioactive changes, known as the actinium series, and it seems probable that this kind of lead is the final product of this series. No other element of the series can be separated to have its atomic weight determined, and as the actinium phenomena occur however carefully the uranium is purified, it is probable, therefore, that this kind of lead is derived from a second kind of uranium with atomic weight 235 or 239. Atoms with the same chemical properties but different masses are called isotopes. The two kinds of lead derived from uranium and thorium were the first examples known, but Aston has found that actually the majority of the chemical elements are mixtures of isotopes. The apparent constancy of atomic weights is due to the occurrence of isotopes together always in the same proportion. His work involves some reconstruction of the atomic theory of chemistry, according to which all atoms of the same element were supposed to be alike. This is not the place to discuss the results, which would lead directly to problems of atomic structure, but it appears that the activity of the actinium series implies an age not greater than 3000 million years.

We can therefore say that the age of the earth lies between 1500 and 3000 million years, and this is consistent with the results of other lines of investigation, though none of the others is so definite. I am inclined to think that tidal friction may give a closer limit when tidal theory is more fully developed; the difficulties of the theory lie in working out the mathematics.

The idea that the earth had a beginning leads us at once to ask where it came from; and here geophysics borders on cosmogony. The older cosmogonists, such as Laplace, were concerned almost entirely with the origin of the solar system, but in the present century the attention of astronomers has become increasingly concentrated on the stars. Even

108 EARTHQUAKES AND MOUNTAINS

Jeans, the leading cosmogonist of the present day, though he has done work of the first importance about the origin of the solar system, has done much more about the structure and development of the universe as a whole. All cosmogonists, even before we were in a position to give an upper limit to the age of the earth, supposed the earth and the other planets to have been formed in some way from the sun, though they disagreed about the method of formation. For purely geophysical purposes, once the solar origin is granted, the details are not of great importance; any matter ejected from a gaseous star must have been gaseous, and we need consider only how the earth passed from a gaseous state to its present one, in which about an eighth of the volume is liquid and the rest solid. But some statement of the main features of the problem is worth while here. The solar origin meets with a great difficulty at the outset. We require the planetary matter to have been ejected from the sun, and this implies some cause making for expansion; but we also require it to have collected again into separate bodies, many of which are denser than the sun, so that we also need an explanation of the subsequent condensation. It is anything but obvious that there is a possible cause of the first process that would not prohibit the second, and it cannot yet be said that the existence of one is established. In the old nebular theory, the sun was supposed to have been distended beyond the orbit of the most distant planet, and to have formed the planets in one way or another because it came to rotate too fast to hold together. The theory can be stated in several forms, but all break down over this essential difficulty, as Jeans and I showed (for different cases) in 1916 and 1918. If the rotation was fast enough to produce disruption, it would also be too fast to permit condensation of the ejected matter into planets. For very large masses there is no inconsistency; a star rotating too fast to hold together can form a double star, and a nebula can form spiral arms capable of condensing into stars, but bodies of the masses of the planets, if formed from the sun, would never condense. In recent years, therefore, more attention has been paid to a biparental theory of the origin, which was actually given before Laplace's time by the French naturalist, Buffon. According to this theory the planets were formed from a splash produced from the sun by the tidal action of a passing star or by actual collision. In either case the rotation would not be so rapid as to prevent condensation later. On the tidal theory the attraction of the star raised two enormous tides in the sun, and it can be shown that if it came too close to the sun the latter would actually break up. (Jeans showed that in such an encounter only the smaller of the masses can be broken up; the tidal theory does not imply that the star developed a solar system of its own.) The matter pulled off would not travel away with the star, for its original velocity was that of the sun; if the star could carry it off it would have carried off the sun too.

The filament thus formed is supposed to have condensed into planets by its own gravitation. If the masses actually collided there would be a thin layer of separation between them, in which there would be rapid shearing with formation of eddies and the other phenomena of turbulence in fluid flow, but the result would again be the formation of a ribbon or filament. This theory has the advantage over the tidal theory that it provides a natural explanation of the rotation of the planets. It is very difficult to see how purely gravitational forces could produce such rotation (we must recall that nearly all the planets rotate much more rapidly than the sun) but the vortical motion in the shearing layer explains rotation quite naturally.*

Either theory has to face the question of how the large differences of velocity within the filament failed to separate its parts so far as to prevent any condensation, especially as the ejection into an almost perfect vacuum would give another reason for expansion. It seems probable that actually much of it did fail to condense. But the expansion led to a rapid fall of temperature, as in a liquid air machine, and the substances with high boiling-points formed liquid drops. This would remove the chief cause making for expansion, and it seems probable that when this condensation had occurred the planets would be able to hold themselves together. The gaseous constituents escaped and were spread through

* *M.N.R.A.S.*, **89**, 636-641, 731-738, 1929.

RADIOACTIVITY AND EARTH'S HISTORY

the system. The theory has not been worked out in detail; the mathematical difficulties are appalling, and the physical principles involved must involve the constitution of the stars, about which at least three different and equally well supported theories exist, and also, in the collision theory, the details of the distribution of velocity in the shearing layer, allowing both for friction and the local pressure. The collision theory, stated only in the rough way that has been possible hitherto, does lead to three quantitative predictions that agree with the facts. It gives estimates of the total mass of the planets, of their angular momentum of rotation, and of the rotation of the sun, all of which are of the right order of magnitude. There is an alternative explanation of the rotation of the sun, but no other theory attempts to explain the other facts. It can at least be said that the collision theory will be very difficult to disprove.

The notion of a filament needs qualification. A long narrow filament would be unstable from the start, and would never be formed. It is simply a stage that might arise if disturbances making for instability were absent, just as we may think of Humpty Dumpty as sitting on the wall and then falling off; he would actually fall off at once, but it is convenient to think of the two stages in turn. The filament would extend a certain way from the sun, become unstable, and break off; another extension would give a further fragment, and so on. It is probable that the satellites were formed at the same time as the planets; no satisfactory theory of the formation of satellites from their primaries exists, and it seems necessary to regard them as splashes formed when the primaries separated from the sun.

The lost gaseous matter played an important part later, for its resistance to the motion of the planets through it served to make their orbits nearly circular. Its own viscosity led to its disappearance; it was gradually reabsorbed into the sun, this process incidentally accounting for a large fraction of the sun's speed of rotation. The time needed for these changes can be estimated roughly; it is of the order of 2000 million years, in satisfactory agreement with the age of the earth found from radioactivity.

These remarks on the origin of the solar system are included only to indicate the direction of modern lines of inquiry. For our present purposes we need only notice that they imply a gaseous earth, rapidly liquefying and collecting into a single mass, and then solidifying as cooling proceeded further by radiation from the surface. The method of solidification, however, needs discussion. The heavy metals would settle quickly to the centre, where they still remain. The rocky shell would cool by what is usually called convection, a rather complicated process, which requires special attention.

Everybody knows that water can be heated more quickly from the bottom than from the top. When it is heated from the top the upper layers expand and become less dense; they can then rest on the lower

RADIOCATIVITY AND EARTH'S HISTORY 113

ones till further notice. Transfer of heat downwards is then by conduction, a slow process. But if the heating is from below, the warm lower layer breaks up, rises towards the top, and mixes with the rest. Thus as fast as new heat is supplied it is redistributed through the body of the fluid. The essential difference between the two cases is that water heated from the top and conducting heat downwards is stable: water heated from the bottom is unstable. If the water was originally exactly at rest, the bottom perfectly plane, and the rate of supply of heat uniformly distributed over the bottom, it would theoretically be possible for it to remain at rest in spite of the heavier water being on top; but the slightest disturbance would produce currents carrying the warm water up and replacing it by colder water descending, to be heated in its turn. The more rapidly the heat is supplied, the stronger are the currents and the more quickly it is redistributed. With the same rate of supply of heat, the fall of temperature with height is less, usually much less, in this convective state than if it were carried up by pure conduction.

This description is too simple in two respects, both of which are amenable to theory. The instability depends, we have seen, on the heated water at the bottom being less dense than that higher up. But when two layers are interchanged each is under the pressure corresponding to the new depth, and the change of pressure itself involves a change of density. If the rising water is not lighter than the water in its

114 EARTHQUAKES AND MOUNTAINS

new position *when it gets there* there will be no instability. This condition is found to imply that instability cannot arise until the rate of decrease of temperature with height reaches a certain amount depending on the thermal properties of the fluid. We do not often have to deal with such a depth of water that the difference of temperature between the top and bottom is conspicuous, but we do in the atmosphere. Dry air can become unstable only if the temperature decreases with height at a rate exceeding 10° C. per kilometre: for air saturated with water vapour the rate is about 6° per kilometre. When the ground is strongly heated a rising current of air is produced; as it rises it cools under the reduced pressure and may reach such a temperature that some of its water has to condense. On rising further more water separates, and the result is a cumulus cloud. This is the explanation of the large woolpack clouds that are so prominent on hot days in summer. The rate of variation of temperature with height, on the supposition that each part of the air receives no heat from its surroundings as it rises or falls, is called the adiabatic lapse rate; and our first principle is that the fluid is stable as long as the lapse rate does not exceed the adiabatic one.

To produce instability it is actually necessary for the lapse rate to exceed the adiabatic one somewhat. We cannot simply interchange two layers by moving one up and the other down: they would get in each other's way. There must be ascending currents in some places and descending ones in others. The rising fluid, when it reaches the top, spreads out horizontally till it reaches a place where the fluid is sinking, and then descends again. This motion is resisted in two ways. It involves distortion and is therefore resisted by viscosity, and conduction between the places of higher and lower temperature at the same level tends to destroy the differences of temperature, which are necessary to maintain the movement. Thus some variation of temperature beyond the adiabatic rate is needed to prevent the currents from being immediately stopped by conduction and viscosity. In a deep layer this excess is usually negligible, but in a shallow one, with high viscosity, it may be very important.* An important practical instance is in the cooking of porridge in a single pan. If it is not stirred the swelling of the grains produces a large increase of the effective viscosity, and it can be shown that a difference of temperature of several hundreds of degrees between the top and bottom may be needed before instability arises. Thus the upper surface may be well below boiling-point while the bottom is charring. Stirring of course destroys this great temperature difference by mixing the layers.

We notice that the existence of the currents requires that they shall have room to pass. If water is heated in a long vertical tube, instead of being in a vessel broad in comparison with its depth, the rising current

* Jeffreys, *Phil. Mag.*, **2**, 833-844, 1926; *Proc. Roy. Soc.*, **118A**, 195-208, 1928; *Proc. Camb. Phil. Soc.*, **26**, 170-172, 1930.

must be on one side and the descending one on the other, and if the tube is narrow the viscous resistance will be much increased. This fact is probably connected with the behaviour of geysers. The water may be unable to conduct the new heat away except at such a temperature gradient that boiling must occur at a certain depth, while viscosity prevents convection. Thus steam is formed internally, with an increase of volume which expels some of the upper water. This produces a relief of pressure which vaporizes more water until the whole has been blown out. New water soaks in from the surrounding rocks and the process is repeated. A proper adjustment of conditions is needed to give the periodicity characteristic of so many geysers, but an explanation on these lines is at least qualitatively possible.

In molten rocks the adiabatic lapse rate is about $0.3°$ per kilometre. So long as the earth was liquid a negligible increase above this would be needed to produce instability, and rapid cooling at the top would take place by radiation. Thus the cooling would be by convection, and the adiabatic lapse rate would be maintained until solidification began. But the melting-point in turn depends on the pressure. In substances that contract on freezing the melting-point is raised by pressure, so that the melting-point at great depths is higher than at the surface. The rise of melting-point with depth at low pressures is about $3°$ per kilometre; it probably decreases somewhat at greater depths, but it is very unlikely to become

RADIOACTIVITY AND EARTH'S HISTORY

less than the liquid adiabatic gradient at any depth. Thus while the earth was liquid the temperature would increase downwards, but the melting-point would increase faster, and as the whole would cool at a uniform rate the melting-point would be reached first at the bottom. The rocky shell therefore solidified from the bottom upwards. The core, however, would remain liquid. At ordinary pressures iron has a lower melting-point than the more basic silicates, and its melting-point is probably lowered by pressure, so that when the lower parts of the shell became solid the core was still far above its melting-point. The core was therefore trapped below a layer of badly conducting silicates and has cooled little further. It should have been anticipated from thermal considerations that the central core would still be liquid, though as a matter of history it was not.

The whole process of solidification probably did not take more than some tens of thousands of years. If there was no further supply of heat we could use the present thermal state to determine the age, and this method was used by Kelvin. We must notice here that the temperature at the surface is controlled almost entirely by solar radiation; it is such that the loss of heat due to radiation into space just balances that received. The average rate of supply of heat from the sun over the surface is 0·007 calorie per square centimetre per second. The rate of conduction from the interior is the product of the conductivity and the temperature gradient, say $0·008 \times 0·00033$ or $2·6 \times 10^{-6}$

calorie per square centimetre per second. In determining the present temperature the internal heat is therefore entirely negligible. If the surface temperature has varied during geological time it is due to meteorological causes and perhaps to changes of solar radiation: internal heat has nothing to do with it. The point needs emphasis, because the contrary opinion was stated as a fact in an important geological text-book published only a few years ago. Cutting off internal heat entirely would only lower the surface temperature by about 0·03°.

The heat being conducted out of the earth is partly a relic of the original heat and partly heat generated more recently by radioactivity. The latter part is the greater. All rocks found at the surface contain small but measurable amounts of uranium and thorium. In addition the common metal potassium is very slightly radioactive. It does not form helium, but one of its constituent isotopes, of atomic weight 41 (ordinary potassium being 39), loses electrons and is converted into an isotope of calcium. These radioactive changes give rise to heat. It is found that uranium, thorium, and potassium all generate heat at much the same rate; the smaller activity of potassium is compensated by its greater abundance. In average granite the total rate of generation of heat is about $1·3 \times 10^{-12}$ cal. per cubic centimetre per second. This does not look much, but when we compare it with the amount of heat being conducted out of the earth we see that the whole of the latter would

RADIOACTIVITY AND EARTH'S HISTORY

be supplied by a layer of granite 2×10^6 cm., or 20 km., thick. This leaves nothing to be supplied by the radioactivity of the rocks below such a layer or by original heat. Thus 20 km. is an upper estimate for the thickness of the granitic layer. The value obtained from the study of near earthquakes and surface waves is about 10 km., with an uncertainty of perhaps 3 km. The whole radioactivity of the earth below 10 km. down, together with the original heat, supplies no more heat to the surface than the top 10 km. do. This leads to the consequence that the radioactivity must decrease rapidly with depth; or, in other words, that in the early history of the earth nearly all its radioactive constituents must have become concentrated near the surface.

It is not very clear how this happened, but there can be no doubt that it does happen in suitable conditions and that we are led to impossible conclusions if we try to evade it. Holmes, for instance, has noticed that in a series of granites from Finland the radium content was multiplied by 2·6, thorium by 6, and potassium by 2 as time went on. Here we have a series of samples at successive dates from the top of a magma, indicating that as time goes on the radioactive elements tend to move upwards. The radioactive generation of heat from average basalt is $0·50 \times 10^{-12}$ cal./cm.^{3} sec.; from dunite, a rock composed almost entirely of olivine, $0·13 \times 10^{-12}$ cal./cm.^{3} sec. On comparing these data with those for granite we see that there is a strong tendency for the radioactive elements to become associated with the more acidic and lighter rocks. But even for dunite, a layer 200 km. thick with the above radioactivity would provide all the heat coming out of the earth, and the whole thickness of the rocky shell is 3000 km. We must suppose that the radioactivity continues to decrease with depth even in the lower layer.

The reason for the upward concentration of potassium is fairly clear. In the cooling of the crust the first compounds to crystallize would be those forming olivine, which are also the densest and would sink. This would continue till the composition of the liquid was so much changed that new constituents began to crystallize, beginning with those containing sodium and calcium, and finishing with those containing potassium. The second stage corresponds to the formation of the intermediate layer and the last to that of the granitic one. For uranium and thorium the explanation is more difficult. These elements and their compounds are very dense and might be expected to sink in a magma: the fact is, however, that they do not. Holmes suggested that they rise because they form volatile compounds, which would be carried up by the steam generated during crystallization. This remains a possible explanation, though it is hardly established. Differences of solubility could also explain the facts.* What seems to be the most probable

* Ether dissolves carbolic acid out of an aqueous solution, although carbolic acid has a density of 1·06; the ethereal solution floats on the aqueous one.

explanation has been given by V. M. Goldschmidt.* Investigation of the structure of crystals by means of X-rays has led to precise determinations of the arrangement of the atoms, which are built up into a regular repeating pattern. In silicates the structure is mainly determined by the oxygen atoms, which are larger than the others; the oxygen atoms are packed as closely as possible, while the silicon and metallic atoms are packed into the interstices where there is room for them, in just sufficient numbers to hold the whole together. For any atom to enter into a silicate it must have the right size to fit into the vacancies left by the oxygen atoms. It appears that the atoms of the radioactive elements are larger than those of magnesium, iron, and aluminium, about the same as those of calcium and sodium, and rather smaller than those of potassium. The metals in olivine are magnesium and iron. It appears that uranium and thorium could be accommodated only in the minerals containing calcium, sodium or potassium, and not even there without some rearrangement of atoms, on account of the difference in valency. A liquid containing a little of these elements and crystallizing will deposit crystals that are quite free from them; they will stay in the solution till the last stage. In actual rocks they are found to be confined to the interstices between the crystals, as we should expect on this theory. In a deep magma the crystals sink to the bottom, leaving the uranium and thorium in

* *Die Naturwissenschaften,* **18**, 999-1013, 1930.

the solution on the top. When the magma is hundreds or thousands of kilometres deep, as in the early history of the earth, the interstices will be closed by the pressure and the solution will be squeezed up to the top. It is probable therefore that below a certain depth there are no radioactive elements at all.

I think that the reduction of radioactivity with depth is generally accepted by writers on the earth's thermal state, but there is a difference of opinion about its amount. If the temperature gradient at the top of the lower layer is, say, 10° per kilometre and supplied by radioactive heating below that level, fusion temperatures would be reached below 200 km. and we should have to regard the whole of the earth below that level as liquid. This on the face of it disagrees with the evidence of seismology, the strength of the shell, and with the height of the bodily tide. It can be shown that if there was any continuous fluid layer within such a distance of the surface the outer surface would be deformed by tidal forces as much as the ocean, and no tides would be observable. Further, if the earth was in such a state there is no obvious reason why the same should not apply to the moon, and the differences between the moments of inertia of the latter should have disappeared. Even this gradient implies that the average radioactivity in the lower layer needed to maintain it is only 2×10^{-15} cal./cm.^{3} sec., $\frac{1}{000}$ of that of granite. In case, therefore, a rapid decrease of radioactivity with depth is required, there is no room for any appreciable radioactive heating at depths beyond 200 km. This conclusion has been contested by Joly, J. H. J. Poole, and Holmes. Joly did not consider any material but basalt to exist below the granitic layer, and assumed for it the radioactivity of surface basalt. He supposed that when fusion temperatures are reached the new heat is used up in supplying the latent heat of fusion, without allowing for the fact that conduction into the solid outer layer is still going on. Other features of his theory have been criticized, and as far as I am aware it has no supporters at the present time. On Poole's theory * a fused layer is formed and gradually melts its way upwards, solidifying at the bottom as it does so; the present state of complete solidity is regarded as a temporary exception. Holmes supposes that there is an absence of strength at all depths and that the new heat is carried up by convection, appealing to the mechanical effects of these permanent convection currents to explain certain geological phenomena. He regards the stress-differences that must exist at great depths as temporary and not as indicating strength. Such an opinion is intrinsically tenable; if the lower layer was a sufficiently viscous fluid moving in a suitable way the stress-differences could exist. But in convection the whole would be thoroughly mixed, the ascending currents would be continually changing their position, and I can see no reason to suppose

* Sci. Proc. Roy. Dublin Soc., 19, 385-408, 1930.

that the average stress-difference at any place over a reasonably long time would differ appreciably from zero. My fundamental objection to all three theories, however, is that they fail to draw the natural inferences from the facts. If we had no evidence on the point we might regard it as equally likely that the radioactive elements are concentrated upwards or uniformly distributed. In the latter case solidification would be impossible and heat would be carried up by convection as fast as it was liberated. We can reconcile the former result with seismology by making the *ad hoc* assumption that the material possesses rigidity without strength; but the whole of the new heat would be carried to the surface, and if radioactivity was uniform the temperature gradient at the surface would be 200 times as great as it actually is. To fit the actual temperature gradient Holmes must assume a great diminution of radioactivity with depth, just as I do; and I cannot reconcile this with the idea that the whole of the lower layer is thoroughly stirred up by convection currents. A full investigation of his theory and of Poole's is desirable, on account of their interest in pure hydrodynamics; but in geophysical applications I think that both are trying to ride simultaneously two horses that insist on travelling in opposite directions.

On my view we must regard the shell as solid and the heat transfer within it as due to conduction. Starting with a state where the temperature everywhere was equal to the melting-point at the corre-

RADIOACTIVITY AND EARTH'S HISTORY

sponding depth, and allowing for radioactive heating near the surface, we have a definite problem to find the distribution of heat at any later time. In a long enough time, if we neglect radioactivity in the lower layer, the temperature in the lower layer and the core would become uniform and equal to that at the surface; but this would take hundreds of thousands of millions of years, and the radioactive elements would have disappeared. The age of the earth is actually short compared with the lives of uranium and thorium, and it is better to suppose that the rate of generation of heat has been constant. We can then estimate the thickness of the upper layers that would give the observed temperature gradient at the surface after the actual time, and infer the cooling at other depths. The gradient is found to be consistent with thicknesses of 11 and 22 km. for the granitic and intermediate layers respectively, which agree with those indicated by seismology. The cooling has, of course, been greatest at the surface; at other depths it consists of two parts. If we ignore the variation of melting-point with depth we find a cooling of about 800° at the top of the lower layer, 200° at a depth of 300 km., and practically none at depths beyond 600 km. But the variation of melting-point with depth implied an original rise of temperature towards the centre; even if there was no radioactivity and the surface was originally cool this would imply conduction outwards and therefore cooling at all depths. The effect is not important near the surface, but it appears that at depths exceeding 600 km. or so there has been a cooling of the order of 50° from this cause. This effect was overlooked until 1932 and must be added to that shown in earlier estimates.*

These estimates refer to average conditions within the continents. The thermal gradient, the conductivity of the surface rocks, and the radioactivity of the local igneous rocks all vary from place to place, and it is very desirable that determinations of all three should be made for the same place as far as possible. The variations may be expected to be associated with the geological history and the local structure, and thermal methods may come, if applied in greater detail than has yet been possible, to give us much more information about these than we yet possess. Determinations of the conductivity of the rocks where the temperature gradients have actually been measured are particularly important to improve our knowledge of the earth's actual heat output and its variation from place to place. The average rate of rise of temperature with depth in North America is about three-quarters, in South Africa about one-half, of what it is in Europe, and we are not yet in a position to say whether this is due to a difference in the thickness of the upper layers, in their radioactivity, or in the conductivity of the upper rocks penetrated by mines. It is on account of the comparatively low temperature gradient that mining operations can be carried out in

* *M.N.R.A.S., Geophys. Suppl.* 3, 6-9, 1932.

South Africa at such great depths; a mine near Johannesburg is 2·5 km. deep, with a temperature at the bottom of 36° C. (97° F.). At a similar depth in Europe the temperature would be about 80° C.*

The temperatures within the earth are very relevant to the cause of volcanoes and other igneous phenomena. The above results were obtained on the supposition that the material of the lower layer would have a melting-point of 1400°; this is rather low for olivine if dry, the normal value being 1500° to 1600°. A basaltic magma would begin to deposit crystals at 1250° and a granitic one at 1000°, according to J. H. L. Vogt. When the lower layer became solid the upper ones still remained fluid on top of it for some time. But the crystallization of the lower layer would drive most of its volatile constituents upwards and into the upper ones; this applies especially to water. The presence of water in the upper layers must reduce the melting-point considerably. Basalt in the crater of Kilauea may remain liquid down to 600°; but when it once solidifies its water is expelled into the atmosphere, and it cannot be melted again below 1200° or so. It seems to be largely owing to the effect of water that igneous activity can occur at all; the

* While this book was in the Press two papers by Dr. E. M. Anderson (*Gerlands Beitrdge*, 42, 133-159, 1934; 43, 1-18, 1934) appeared, suggesting that I have over-estimated the mean conductivity in Europe and the mean radioactivity of the granitic layer; his values are about $\frac{2}{3}$ of mine. He also shows that the residual temperature gradient left by the glacial period may be appreciable. The effect on the calculated thickness of the radioactive layer is not great, but his work adds considerable force to the above remarks.

calculated temperatures at all depths in the upper layers at the present time are well below the melting-points of the dry rocks, but not much, if at all, below those of the same rocks if they contain much water. R. W. Goranson has shown * that under a pressure of about 4×10^9 dynes/cm^2., corresponding to a depth of 15 km., a granite magma can hold 9 per cent. by weight of water in solution; at a quarter of this pressure it can hold 6 per cent., and its melting-point is only 720°. E. M. Anderson and E. G. Radley have found glassy inclusions in an igneous rock in Mull, where crystallization seems to have been prevented because the water was unable to escape.† These facts suggest that much of the water in the upper and intermediate layers has been held in them by the pressure, and that they have consequently failed to crystallize. In such conditions the cooling would leave the rocks in a glassy state. This is contrary to the usual geological opinion, which holds that these deeply buried rocks would cool slowly and form large crystals; but if the retention of water would prevent crystallization this argument loses its force. At present it seems quite likely that the intermediate layer is still glassy and identifiable with glassy basalt or tachylyte. It is certain that lava arriving at the surface contains enough volatile constituents to lower its melting-point by several hundred degrees.

* *Amer. J. Sci.*, **22**, 481-502, 1931; **23**, 227-236, 1932.
† *Q.J. Geol. Soc.*, **71**, 205-217, 1915. I am indebted to Dr. L. Hawkes for the reference.

RADIOACTIVITY AND EARTH'S HISTORY

The calculated temperature at the base of the intermediate layer is 650°, not much below the probable fusion point. Any increase of the temperature due to chemical action, a temporary but occasionally important phenomenon, or to an excess of radioactivity, would lead to fusion. The resulting basaltic magma would then melt its way upwards by convection, aided by the reduction of density associated with melting. This agrees with the fact that the more basic magmas, which would be expected to have come from the greater depths, are usually the first to appear at the surface; though crystallization during intrusion alters the composition and complicates matters.

The water expelled at the surface naturally finds its way into the ocean, the total volume of which must be increasing. When the earth was fluid at the surface it could have little atmosphere and no ocean. It is probable that most of the ocean has come out of the earth during geological time. A fact noticed by Aston supports this view. The amounts of the inert gases in the earth and the atmosphere together are of the order of a millionth of those of even the rarer of the other elements. This is explained at once if the earth was originally heated; for these gases, being unable to form compounds, would have no possible resting place except in the atmosphere and would be lost by diffusion into space. The materials of the present atmosphere and ocean, being volatile and having molecular weights less than those of krypton and xenon, would also be lost, and must therefore have been enclosed in compounds within the earth.* The water and carbon dioxide have presumably been expelled as such, and most of the carbon dioxide has been converted into oxygen and organic compounds by plants. Goldschmidt † has noticed that the present oxygen is about enough to combine with all the organic remains; and the atmosphere of Venus appears to consist chiefly of carbon dioxide. These facts agree with our hypothesis and indicate that there is no plant life on Venus. The nitrogen of the atmosphere was probably expelled as ammonia, which is an important constituent of the atmospheres of the great planets.‡

Meteorites are found to consist chiefly of silicates, a nickel-iron alloy, and of sulphides, principally ferrous sulphide. The silicates are analogous to the earth's rocky shell and the metallic alloy to the core. If we can regard meteorites as a sample of the kind of material that the earth was made from we should expect also that the earth would contain a layer of sulphides. The amount of meteoric accretion is so small that we cannot suppose that the earth has grown much from this cause; the present rate would give a deposit a small fraction of a millimetre thick in the whole age of the earth.§ But it is possible that both the earth and meteorites are specimens of some other

* Jeffreys, *Nature*, **114**, 934, 1924.
† *Fortschr. d. Mineralogie*, **17**, 112, 1933.
‡ R. Wildt, *Göttinger Nachr.*, 171-180, 1932.
§ Jeffreys, *Nature*, **132**, 934, 1933.

body. Goldschmidt points out that in the liquid state the alloy, the sulphides, and the silicates would not mix, and expects therefore that there would be a thick layer of sulphides within the earth. Geophysical investigation has not revealed such a layer, but can say where it must be if it exists. The density of ferrous sulphide is nearly 5, and if such a density existed near the outside the moment of inertia would be too high. Seismic waves show no sign of a sudden change of velocity at any depth between 300 km. and 3000 km., so that the whole of the shell must be formed of silicates. The sulphides, if they are important, must therefore form the upper part of the core. They are more fusible than iron, and would therefore be liquid. Present seismic data will not separate such a sulphide layer from the rest of the core. They can determine only the velocity of a wave at the deepest point of a ray, and all the observable core waves have penetrated to a considerable depth.

Recent work by K. E. Bullen, myself, and Miss I. Lehmann has shown that there is a change in the nature of the P wave at a distance of about $20°$, such as would correspond to an increase of velocity of about 10 per cent. at a depth of about 300 km.; the quantities are provisional and will need revision when the calculations are completed. The nature of this discontinuity is not yet explained.

The early history of the earth must apparently be regarded as the record of the stages of crystallization.

132   EARTHQUAKES AND MOUNTAINS

The first substances to crystallize would sink to the bottom, and this would continue till the extraction of these had so altered the constitution of the mother liquor that it was saturated in respect of some new constituent. This would then proceed to separate until a third stage entered; the upper layers are to be regarded as the record of these successive stages. It is here that geophysics makes contact with the chemical side of petrology. A great deal is now known about the physical chemistry of the silicates, and it should become possible to find out in the near future how thick the layers of transition between the fairly uniform layers indicated by seismology should be. Separation in this manner does occur in igneous intrusion, but more work needs to be done before we can test its quantitative application to geophysics.

Special attention will probably have to be paid to the problem of eclogite. This is a rock with the same general chemical composition as basalt, but consists of a different set of chemical compounds with higher densities. It appears to be formed from basalt under high temperature and pressure. Consequently it was suggested by Fermor that in the deeper parts of the crust basalt would normally be replaced by eclogite, and many geologists are disposed to believe that the greater part of the rocky shell is actually eclogite. I do not personally think that this is likely. Eclogite is a mixed crystalline rock, and a magma with the general composition of eclogite would separate in crystallization into layers; I cannot conceive an eclogite layer reaching nearly half-way to the centre. Laboratory tests on the compressibility of eclogite indicate a velocity of the P waves rather too high for the upper part of the lower layer, which does seem to fit olivine. Now the constituents of eclogite contain aluminium; olivine does not. It would be tempting to suppose that the lower layer down to 300 km. is olivine, with eclogite below that level, but it is improbable that we could have a layer free from aluminium separating the upper layers from a deep layer, all containing it as an essential constituent. A high-pressure modification of olivine seems a more likely constitution for the matter between a depth of 300 km. and the core.

CHAPTER VI

THE BODILY TIDE AND TIDAL FRICTION

" When heaven itself goes out of its way to set a correcting omen in the sky, who dare disobey ? "
—*Kai Lung's Golden Hours*, 116.

TIDES arise from the fact that the earth has a finite size; the gravitational attraction of the sun and moon is not the same at all points of it. The nearer parts are attracted more strongly than the centre, the more remote ones less strongly. The total force produces the acceleration of the centre; the part due to the sun maintains the earth's motion about the sun, while that due to the moon maintains a monthly revolution about the centre of mass of the earth and moon together. These motions would be just the same if the earth was a rigid body. But in the actual earth the excessive attraction on the near side and the deficiency on the far side tend to stretch the earth along the line of centres; these constitute the tidal forces. They can be represented by what is called a gravitational potential corresponding to a spherical harmonic of the second order: that is, the level surfaces due to the earth and the tidal forces together are ellipsoids. The elevation of the level surface due to the tidal force alone is easily calculated, and is called the equilibrium tide; it is not the actual

BODILY TIDE AND TIDAL FRICTION 135

tide, but is a convenient intermediary for theoretical purposes. If the ocean covered the whole earth, and had a negligible density, and the sun or moon remained in the same place with reference to the earth's surface, the ocean surface would settle into such a position that its height at any place was equal to the equilibrium tide. In fact, all of these postulates are untrue, but the equilibrium tide serves as so useful a standard of comparison that it occupies a prominent position in all theoretical work.

If the body of the earth was fluid, and the tidal forces constant, the whole of it would be distorted in such a way that each layer of equal density was a level surface. The upper and lower surfaces of the ocean would be raised or lowered by practically the same amount, and there would be no visible tides in the ocean. The existence of tides is evidence that the earth has rigidity. On account of the irregular form of the ocean and the difficulty of carrying out tidal observations except on the shore it is not possible at present to work out the tidal theory for the whole ocean, but useful results can be obtained from observations of water in long pipes, where we can measure the variation of level at the ends, and from enclosed bodies of water such as the Red Sea and Lake Baikal.*
It appears that the tidal elevation is about 0·6 of the equilibrium tide. Other information is given by what is known as the 14-monthly variation of latitude. A rigid body could rotate permanently about any of

* S. F. Grace, *M.N.R.A.S., Geophys. Suppl.* 2, 301-318, 1931.

three perpendicular axes at its centre of mass, which are called its principal axes of inertia. But if it is set rotating about any other axis, the axis of rotation is not fixed in the body. For an oblate spheroid rotating about an axis near its axis of figure, the axis of rotation revolves around the axis of figure in a period $A/(C - A)$ times the period of rotation, where C is the moment of inertia about the polar diameter and A that about a diameter in the plane of the equator. This ratio is known for the earth from the rate of precession, and gives a period of 305 days. This should be observable, because latitude is measured as the inclination of the horizon to the axis of rotation, and if the axis of rotation is moving in the earth the latitudes as measured should vary in this period. This result was predicted by Euler, but the observed period is about 430 days instead of 305. The difference was explained by Newcomb as due to the earth's being elastic instead of perfectly rigid. The earth's ellipticity can be regarded as composed of two parts. If the rotation was stopped and the earth continued to obey the laws of elasticity, the ellipticity would sink by the amount due to the rotation, but a part would remain. In the actual motion the former part corresponds to a symmetrical bulge about the axis of rotation, the latter to one about a fixed axis, and it is only the latter that affects the variation of latitude. Thus the period of the variation of latitude gives further information about the earth's elasticity.

These two data together were shown by Love to

BODILY TIDE AND TIDAL FRICTION

give an actual determination of the earth's tidal yielding; the elevation of its surface is 0·6 of the equilibrium tide. From this we can find what rigidity the earth would have to have if it was uniform, in order that it might yield by this amount. The result is slightly more than the rigidity of steel. This seemed so surprisingly large when it was first obtained (by Kelvin) that it was generally believed for a long time that the whole of the earth must be extraordinarily stiff. But the results of seismology and the theory of the figure of the earth have now shown that the lower layer at its top is about two-thirds as rigid as steel, and the base of the rocky shell more than twice as rigid. The average rigidity of the shell is considerably more than that of steel. Further, on account of the concentration of mass in the core, the tidal forces per unit volume are less than for the uniform earth. We find now that there is no room for any rigidity in the core; if the core had any appreciable rigidity the tidal yielding would be less than it is. This is the decisive evidence that the core is liquid, and not merely a solid that admits distortional waves but damps them out before they have passed through.*

The maximum tidal bulges produced by the moon in the earth would be exactly under and opposite to the moon if the earth was perfectly elastic or perfectly fluid; in a deep ocean covering the whole earth to a uniform depth the same would apply if we could neglect viscosity. Actually there must be

* Jeffreys, $M.N.R.A.S.$, $Geophys. Suppl.$ 1, 371-383, 1926; L. Rosenhead, 2, 171-196, 1929.

some imperfection of elasticity in any real material, and there is no such thing as a non-viscous liquid. There is therefore a continual loss of energy in the tides, which has to be supplied from somewhere. It appears as a resistance to the rotation of the earth. The actual high tide at a place is somewhat after the moon has reached the meridian, so that the tides are out of alignment with the moon, and the moon's attraction on them tends to turn the earth in the opposite sense to its rotation. Thus the earth's rotation becomes slower. At the same time the reaction on the moon tends to pull the moon forward. Rather surprisingly at first sight, the result is to make the moon go further off and revolve more slowly. The reason is that the forward pull on the moon increases its angular momentum about the earth; but the angular momentum is the product of the mass, the orbital velocity, and the distance. The velocity varies inversely as the square root of the distance, so that the angular momentum increases as the square root of the distance. If the angular momentum increases, so therefore must the distance.

The tides in the actual ocean are much more complicated on account of the irregularities of form and depth. In some places high tide comes after, in others before, the moon's meridian passage. This is not a matter of friction; it would happen even in a frictionless ocean where the tidal wave has different velocities in different places on account of the differences of depth. But so long as there was no dissipation of

BODILY TIDE AND TIDAL FRICTION 139

energy the attraction of the moon on the tide would give no resultant couple tending to resist rotation; in places the tendency would be to accelerate rotation, in others to resist it, and on the whole the effect would just cancel. These phase-differences are so large that the effect of friction cannot be disentangled except in limited regions. We can, however, estimate the loss of energy by two methods due to Professor G. I. Taylor. The frictional force in a liquid flowing over a solid is about $0\cdot002\ \rho v^2$ per unit area, where ρ is the density, v the velocity of slip, and $0\cdot002$ is a numerical constant. The formula is empirical and applies only when the motion is turbulent; that is, in nearly all practical cases, including wind and the flow of water through pipes and channels. The rate of dissipation of energy per unit area is therefore $0\cdot002\rho v^3$ taken without regard to sign. Given the velocities in tidal currents we can therefore find the dissipation of energy. The other method is to consider a shallow sea. The tide in such a sea is produced by the pressure on its oceanward side due to the tide in the open ocean; variations in the level of the ocean push the water of the sea landwards and allow it to flow back when the ocean level falls again. We can find the rate of performance of work on the sea by the pressure of the ocean, and as the energy of the sea is not steadily increasing or decreasing the rate of dissipation within the sea must be equal to this rate. The two methods give consistent results.

It appears that in the main oceans the tidal currents

are too gentle to give any important dissipation of energy. It is only in shallow seas around the coasts that the currents are so magnified as to make the dissipation matter. The reaction on the tides in the main oceans must, however, be important, for it seems that the dissipation in the shallow seas is enough to destroy the tidal energy present in the ocean at any one moment within about a day. The tide may be regarded as generated in mid-ocean by the attraction of the moon and spreading out as a wave; when it reaches the coast most of its energy is absorbed by the friction. This resistance affects the phase of the reflected wave and therefore alters the tides over the whole ocean.

Knowing the rate of dissipation we can find how the earth's rotation and the moon's distance are altering. The observable effects are an apparent steady acceleration of the sun and moon in longitude, arising from the fact that the earth's rotation, which is altering, gives us our practical standard of time. This can be detected from observations of ancient eclipses and occultations of stars by the moon. The amount agrees as well as can be expected with that inferred from the dissipation in shallow seas, so that the theory is adequately checked. The earth's rotation is at present becoming slower at such a rate that the day, 120,000 years ago, was 1 second shorter than it is now.

There is no reason, therefore, to suppose that any large fraction of the dissipation of energy in the tides is in the body of the earth. Some of it must be, but not more than a fifth of the whole and probably

BODILY TIDE AND TIDAL FRICTION

a good deal less. For this reason Sir G. H. Darwin's theory of the process needs re-examination. He took the friction to be entirely due to viscosity in the solid earth, and constructed a monumental theory of the effects not only on the periods of rotation and revolution, but also on the eccentricity of the moon's orbit, and the inclinations of the equator and the moon's orbit to the ecliptic. Tracing the changes back to an early time, he showed how the moon can have receded from a position so close to the earth that it revolved and the earth rotated in about 5 of our present hours, the original eccentricity and the inclination of the moon's orbit to the equator being insignificant. We do not know yet how the transfer of the friction to the shallow seas affects the results for the eccentricity and the inclinations; the variation of the periods is hardly affected, because it depends on more fundamental considerations of energy and angular momentum, which will hold wherever the dissipation occurs.

On account of the smaller distance of the moon in the past the tides were larger and the changes more rapid. It does not seem likely that there has been much change during the last 1000 million years or so, but the whole of the changes must have been confined to something of the order of 4000 million years at the outside. Uncertainties in tidal theory and in the past distribution of shallow seas make an accurate estimate impossible at present; but as the non-dissipative forces are proportional to the current and the dissipative ones to the square of the current it seems likely that a good approximation for the remote past will be found by supposing the dissipation to be as rapid as possible.

The former proximity of the moon to the earth leads naturally to the suggestion that the moon was once part of the earth. The combined body would rotate in about 4 hours. This would make the period of the tide raised by the sun at any one place equal to 2 hours, which is about the period of the free oscillation that would ensue if a fluid mass was distorted into an ellipsoidal form and left to itself. We should therefore have the effect known as resonance, which is now generally familiar in the reception of wireless waves. It seemed likely that the solar tide would be so much magnified in these conditions that the mass would be so elongated as to become unstable and break up into two. This is the resonance theory of the origin of the moon. It has had a somewhat chequered career, having at several times met with apparently fatal objections which have afterwards been overcome, but at present the evidence is against it. The friction at the boundary of the core would dissipate the energy of the tide so rapidly that it could never reach an amplitude more than about $1/20$ of the radius, which would be far too small to give instability;* and it seems, according to Nölke, that the velocity of the detached mass, if any, would be so small that it would at once fall back into the primary. The rotation would not be rapid enough to produce disruption without assistance, and it seems

* *M.N.R.A.S.*, **91**, 169-173, 1930.

BODILY TIDE AND TIDAL FRICTION 143

that we cannot suppose the moon to have been part of the earth at any time since the latter had a distinct individuality. It is probably as old as the earth and an immediate result of the same process as gave birth to the other planets and satellites.

The effect of tidal friction on the moon is even more conspicuous than on the earth. The tendency of friction is to destroy the difference between the rate of rotation of the body and the rate of revolution of the tide-raising body about it. For the earth this will not happen until the moon's distance is 1·2 times what it is now. But the tides raised in the moon by the earth would be much more effective in bringing the moon's rotation into agreement with its period of revolution, and this is why the moon keeps the same face always towards the earth. In the same way Mercury, and probably Venus, keep the same face always towards the sun, and all the other satellites whose rotations have been measured keep the same faces towards their primaries. It is unlikely that any of these bodies have ever had shallow seas on their surfaces analogous to those on the earth, so that the friction must have been internal. It should be possible to get lower limits to its amount, which would be interesting, because the study of the earth by itself gives nothing but upper limits. The imperfection of elasticity is probably elastic after working in all cases: there is some evidence of damping in seismic waves that can be attributed to this cause, but it can equally well be attributed to scattering of the waves at interfaces and their ultimate absorption in the core.

CHAPTER VII

THE MECHANICS OF GEOLOGY

"' The matter is as long as The Wall and as deep as seven wells,' grumbled Sheng-Yin, ' and the Hoang-Ho in flood is limpid by its side.' "
—*Kai Lung's Golden Hours*, 115.

WE have already considered the strength needed to support continents and mountains; but this tells us only something about their present state. If we ask also how they have come into existence we shall want an explanation of the stresses needed to produce them. The elevation of a mountain system represents work done against gravity. Further, the stresses, so far as they are frictional, simply resist a change of form. At present the strength of the crust is preventing gravity from making the surface level; but when the mountains were being formed it was aiding gravity in resisting the stresses that made them. To explain the origin of mountains we must provide such stresses as will overcome both the strength of the earth and gravity, for in a symmetrical body both would act together in opposing any change of shape. The only agency that seems capable of supplying such stresses is contraction of the interior. Any internal contraction means that the outer portions either have to stay behind unsupported except by their own internal stresses, or must sink so as to remain supported on the interior. It is easy to show that when gravity is taken into account the latter alternative is correct; an unsupported crust would fracture at once under its own weight, and in any case contraction would not give the type of stress needed to detach a spherical shell from the interior. But a sinking crust has to acquire a shorter circumference to fit the new size of the interior, and this implies an extra pressure across vertical planes to shorten it. The normal stress over the outside is only atmospheric pressure, the variation of which is negligible, so that the contraction of the interior will produce a set of stress-differences at the outside such that the pressure across vertical planes exceeds that across horizontal ones. When these become great enough fracture or flow will begin. These stress-differences will be world-wide, and yield at any point will increase those in its neighbourhood, so that when yield begins readjustment will proceed until several great circles have been shortened enough to bring the whole of the stress-differences again below the strength. The shortening needed can be estimated from what we know already. Young's modulus for granite is about 8×10^{11} dynes/cm.^{2}, and the strength is about 10^9 dynes/cm.^{2} Thus granite can be compressed by about $1/800$ of its length before it is crushed. The circumference of the earth being 40,000 km., the shortening needed to give fracture is about 50 km., and when the outer crust has accommodated itself to this by yield the yield will stop. Any further internal contraction will then accumulate until the crustal stresses have become enough to produce a new yield. We should therefore expect the visible readjustment to occur intermittently, each stage corresponding to a shortening of the crust by about 50 km.

Now geology shows that the great mountain systems have been elevated at intervals with long intermediate stages of quiescence. Also the great epochs of mountain formation have been world-wide. The three most important ones in Post-Cambrian time are known as Caledonian (Silurian), Hercynian (Permo-Carboniferous), and Alpine (Tertiary). The Caledonian system can be traced in Wales, Scotland, Scandinavia, Siberia, Africa, Australia, and South America. The Hercynian movement accounts for many of the mountains of Great Britain, Germany, the Urals, and the Appalachians. The Alpine system includes most of the present great ranges, from the Pyrenees to the Himalayas, the Rockies, and the Andes. The earlier systems are harder to trace as continuous belts of mountains, because they have been heavily denuded and can often be recognized only by special attention to the folding involved in them. It is therefore likely that future work will make their mapping much more complete. There are also signs of three movements of comparable intensity in the Pre-Cambrian. It is therefore natural to identify the stages of crustal yield with the successive great periods of mountain formation.

This inference can be checked by comparing the estimated crustal shortening with that indicated by the mountains themselves. The latter can be found from considerations of isostasy. The present great systems are all compensated, and it is reasonable to assume that gravity was normal before they were formed; in other words, that the formation has not disturbed the uniformity of distribution of mass. The place where mountain formation began would be determined by some local weakness, possibly quite trifling. But the crustal shortening implies that an extra thickness of the upper layers is piled up in one place. To re-establish isostasy an equivalent mass of the lower layer must be pressed out, but its thickness would be less than that of the lighter upper layers, and an elevation would remain. From the amount of this elevation we can estimate the crustal shortening. Suppose, for instance, that the upper layers consist of 1 km. of sedimentary rocks of density 2·4, 10 km. of granite of density 2·6, and 20 km. of tachylyte of density 2·9. If the intermediate layers are not tachylyte 2·9 will still be near their mean density. Suppose also that the mountain formation doubles all these thicknesses. Then the thickness of the upper layers is increased by 31 km. But if the lower layer has density 3·3 the restoration of isostasy will expel 26·2 km. of the lower layer, and the remaining elevation is 4·8 km. For smaller elevations the thickening can be found by proportion. The heights of the mountains therefore provide a means of finding the amount of the compression that produced them.

The heights, however, are not now the same as when the mountains were first formed. The steep slopes that determine the appearance of mountain scenery are not original; they have been carved out by erosion since the mountains were formed. Material has been removed from the valleys to depths of perhaps several kilometres, while little has come from the mountain tops. But the removal of surface rocks, mainly sedimentary, requires an inflow below to restore isostasy, and this raises the whole region. The mountain tops, on account of this secondary effect, are higher now than when the ranges were first formed. This elevation can be estimated from the known depths of the valleys. Allowing for this effect we find that the primitive height of the Alps and Rockies was about 3·2 km., and that of the Himalayas 5·8 km. In this early stage the mountain ranges would be continuous ridges with no valleys intersecting them.

A verification of this argument has been obtained by L. R. Wager from a study of the Arun River, which rises on the Tibetan Plateau at a height of 22,000 feet, flows east for some distance, and then turns south to cut through the Himalayan mass between Everest (29,000 feet) and Kangchenjunga (27,600 feet), forming an enormous gorge. This mass was certainly originally continuous, and for the river

THE MECHANICS OF GEOLOGY

to cross it at its present height it would have had to flow uphill. But the course could have been formed if the original height was only 16,000 feet (4·9 km.), the present height of the southern part of Tibet. Within the main mass the volumes above and below this level are nearly equal. Wager therefore infers that the lightening of the crust due to denudation has led to a general uplift of the Himalayas, which were originally merely the southern edge of the plateau, by 11,000 or 12,000 feet. A possible alternative explanation might be that the river originally rose on the southern slopes of the mountains and has cut back through them, but Wager's study of the sediments laid down by it in its upper course confirms the view that the mountains were rising after the river took its present course.

The primitive height leads, therefore, to an estimate of the average thickening of the upper layers during the formation of the ranges, and therefore to the horizontal shortening that was needed to give such a thickening over the actual width of the ranges. For the Alps and Rockies the shortening needed is about 70 km., for the Himalayas about 190 km. Wager's values suggest that the last may need some reduction. The values for the Alps and Rockies are in good enough agreement with what we inferred from the strength.

The Himalayas present a difficulty not only by themselves but still more when we consider also the plateau of Tibet and the Kuen Lun mountains on its northern margin. The Tian Shan and Altai mountains still further north are of Hercynian age, and therefore do not represent part of the same movement. If the whole of Tibet, with its marginal ranges, was elevated at once, we should require a shortening of about 1000 km. The contraction of the interior would be expected to shorten all great circles by about the same amount, and we have no explanation of how such a large amount can be concentrated in a limited region. It seems that the Tibetan region is abnormal, but whether the abnormality consists in a difference of local structure that would give much more elevation with the same shortening, or to some local effect that gives extra shortening, or whether the elevation of Tibet is due to some other cause altogether, we are not in a position to say. Evidence from seismology or gravity might help, but at present we have none. Apart from this region the shortening required is consistent with theory.

Geologists have obtained independent estimates of the contraction from study of the displacements within the mountains themselves. These are on the whole much more than the geophysical estimates; values of about 300 km. have been obtained for the Alps and the Appalachian mountains. These are found by mapping the folded strata and finding how much they would have to be spread out to become flat again, and by estimating the displacements in thrusts, where one mass of rock has been pushed bodily over another. But it is dangerous to identify the movement measured

THE MECHANICS OF GEOLOGY 151

in this way with the shortening of the crust. A pack of cards can be made to slide over one another to any extent without distorting the table. A well-known geological phenomenon called "hill-creep" is even more pertinent. Sediments resting on a slope show a tendency to slip downwards, and in doing so they develop little fractures and folds closely resembling on a small scale the great ones to be found in mountains. The underlying firm rocks remain unaffected. We cannot therefore assume without direct evidence that the movement shown by folds and thrusts is anything but a superficial one; the displacements below the visible surface may be very much less.

When we come to consider more in detail what would happen when the crust yields to a general compression we find that such shallow movement is to be expected. When a solid is broken the fracture may follow either of two directions. Under tension it may break straight across, but this is not our case. Under compression the plane of fracture is such as to include the greatest and least of the principal stresses, usually at about 30° to the direction of greatest compression.* Yield takes place by the material on one side sliding over this plane. The simplest possible adjustment to a general compression would be as follows. A fracture is formed at 30° to the surface; sliding would proceed along this till the horizontal displacement was about 70 km. and the corresponding

* On the usual Coulomb-Hopkins theory the angle is 45°, but 30° agrees better with experiment. The difference has been explained by E. M. Anderson, *Trans. Geol. Soc. Edin.*, **8**, 387-402, 1905.

vertical one about 40 km., producing an enormous overhanging fault. Isostatic redistribution of load would proceed to give outflow below, but would do nothing to disturb the inequality of the distribution of load at the surface. To support such a precipice we should need a strength about ten times that of granite. It would therefore fracture under its own weight about ten times while it was being formed. It may even seem that, as the stress-differences produced by the elevation are much more than before, the last stage is worse than the first, and that fracture does nothing to relieve the stresses. This, however, is not the case. It has increased the stresses in the immediate neighbourhood, but it has removed them everywhere else. If it were not for gravity it would also have removed the local ones. The stress-differences arising during fracture are then a secondary effect due to gravity, are local, and can be relieved by local movements. But as they would be so much greater than the strength if the elevation reached 40 km. they will produce their own type of yield before the elevation is complete. If the yield was always fracture the edges would break off as they rose and the result would resemble a load of bricks newly turned out of a cart, broken and overturned in all directions. Further elevation would give new fractures, rocks always sliding or rolling from the back of the cliff over the heap already formed. This result resembles the facts sufficiently to be probably part of the truth, though it is not the whole.

CONTINUOUS DISTORTION AND FRACTURE COMBINED,
BROADHAVEN, PEMBROKESHIRE

FOLDING IN SHALE DUE TO HILL-CREEP,
MATLOCK, DERBYSHIRE

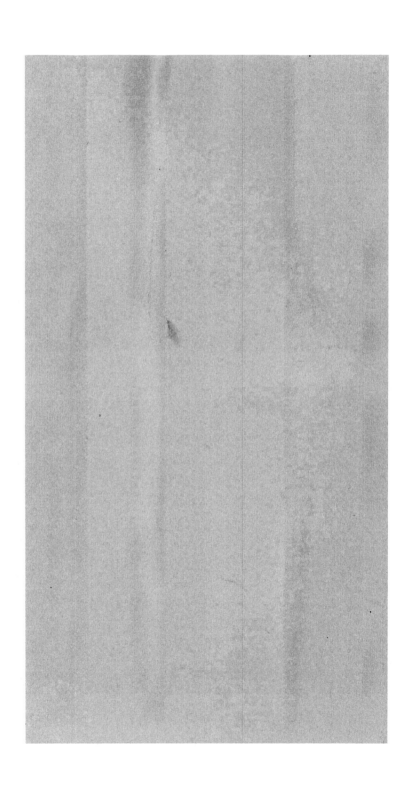

THE MECHANICS OF GEOLOGY 153

Blocks of rocks overturned in all directions are characteristic of mountain systems, and it is found that the rocks elevated later have proceeded from the rear, travelled over those already in their final place, and come to rest on the undisturbed ground in front. The chief difference between this account and the facts is that in great mountain systems the displacements are chiefly continuous distortions; fractures are a regular feature, but are subsidiary to the great folds. The theory, however, provides equally well for folding. We have already noticed that under high pressure or temperature, or both, yield tends to take the form of continuous distortion rather than fracture, and we are thinking here of rocks that have been elevated from a considerable depth and are still under a heavy load and hot. Fracture would therefore be confined to small depths; at greater depths we should have continuous distortion. The details of the changes are less clear than for fracture, but we shall still expect a strong elevation over a narrow belt, and subsequent outflow due to gravity. This will take the form first of a fold, matter flowing from the elevated tract and coming to rest on the low ground in front (called the foreland). Further elevation will give another fold, the matter coming from behind the root of the first fold, but spreading over it and finally coming to rest in front of it. Several such folds may be formed in turn. This structure was first recognized in the Alps by Schardt, in 1894, and has since become the foundation of Alpine geology

154 EARTHQUAKES AND MOUNTAINS

in the hands of Heim, Collet, Staub, and many others. Professor E. B. Bailey has taken a leading part in demonstrating a similar structure in the Highlands

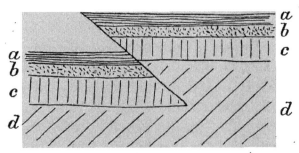

Fig. 6.—Result of crustal shortening by oblique fracture. *aa*, sediments; *bb*, granite layer; *cc*, intermediate layer; *dd*, lower layer.

of Scotland. But most of these writers have taken it for granted that the displacements are the direct result of horizontal pressures acting directly on the rocks involved; these are supposed to have pushed the rocks over the foreland and over the earlier nappes.

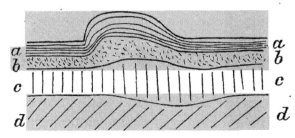

Fig. 7.—Result of crustal shortening by continuous yield.

It is very hard to see what could produce such stresses, apparently confined to a depth of a few kilometres, and two other phenomena argue against

THE MECHANICS OF GEOLOGY 155

their existence: klippes and back-folding. A klippe is a piece of a nappe found detached from the main nappe but far in front of it. If the thrust had to be transmitted through the nappe it could never have

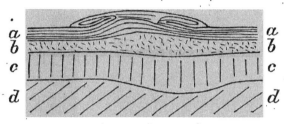

FIG. 8.—Secondary result of crustal shortening, with outflow above due to gravity.

pushed such a piece to a position that the nappe never reached. But if the sliding is a downhill slip due to gravity it is quite intelligible that the nappe might break and that its front portion would move

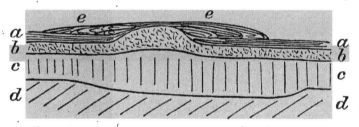

FIG. 9.—Ultimate result of outflow. *ee*, nappes. The symmetry is probably exaggerated.

on independently for some distance. Back-folding is shown by folds with their fronts in the opposite direction to the main nappes. On the theory of a general horizontal movement these rocks would have had to move against the main thrust. But if there is a primary elevation along a narrow belt the resulting outflow might well be outwards on both sides. The amount of back-folding in comparison with the main nappes would depend on the degree of asymmetry in the primary elevation.* If this was a single fracture at a slope of 30° the tendency to forward rather than backward folding would be overwhelming, but in actual conditions the asymmetry may well be less. We cannot yet follow out the changes in detail because they depend on the displacements in materials at stresses above the strength, and at present we have only the outline of a theory. It does appear, however, that the idea of a primary elevation with successive stages of outflow due to gravity is in general agreement with the geological structure of mountains. But while the flow is going on near the surface other flow must be happening down below to maintain isostasy. We cannot say which would happen fastest. When the stress-differences exceed the strength the plastic flow seems to follow the same law as for a viscous liquid, but we do not know the viscosities. All we can be sure of is that both flows will proceed till the stress-difference is everywhere below the strength, and that from that time any further deformation will be elastic unless the strength is again exceeded. The distribution of viscosity with depth, when flow is taking

* Figs. 7, 8, 9, show more back-folding than usually appears to occur. But the number of theoretical possibilities is so large that I have not thought it worth while to try to depict them all, and it seems that all of them can be matched in reality!

place, is our next objective. For stresses below the strength the viscosity is effectively infinite; this is merely another way of saying that plastic flow does not occur at such stresses.

The fracture that leads to a mountain system would be expected to start at the weakest part of the crust. Geological observation shows that mountain formation regularly takes place along regions where sediments have been deposited to a great depth and for a long time. It appears that most of these places were "geosynclines", that is, long and rather narrow arms of the sea, which were gradually filled up by the sediments from the neighbouring land. We seem to have few geosynclines at the present day. The Mediterranean and the North-west Passage approximate to the geosynclinal form, but the Alpine system of mountains, for instance, corresponds to an ancient sea stretching from the Pyrenees to China. Now such deposition provides in itself a source of weakness. It means that any heat rising to the surface has to penetrate an extra thickness of material, so that the sediments act as a blanket. Higher temperatures are therefore to be expected below regions of deposition than elsewhere, and the heating makes for weakness. The association of thick sediments with mountain formation is therefore natural.

The blanketing effect provides an explanation of another geological phenomenon and a partial explanation of a third. At the base of 10 km. of sediments the temperature would rise by about 250°

in about 100 million years.* This combination of temperature and pressure is a necessary condition for the formation of the iron garnet, which is characteristic of sediments that have been buried to such a depth. This thermal metamorphism has been known for a long time, and has been used by Elles and Tilley in the Highlands to determine the original order of deposition where the rocks have in many places been completely overturned by nappe formation. The direct effect of pressure in raising the temperature is much smaller, only about 1°. The sediments themselves would be cold when first deposited; the problem is to explain how they became heated, and it turns out that conduction into their base is about adequate for the purpose.

The other phenomenon is connected with isostasy. On the face of it a depth of 10 km. of sediments appears to imply that the sea was 10 km. deep at least. With any smaller depth the sea would be filled up before so much sediment had accumulated. Actually these deep deposits show in most cases an alternation of limestones, shales, and sandstones or rocks formed from these by metamorphism. There is no sign that the sea became much shallower during the process; on the contrary, the depth seems to have oscillated somewhat through a limited range, but without much systematic change in either direction. To understand these thick sediments we need to know how 10 km. of them can have been deposited in a sea that was

* *M.N.R.A.S., Geophys. Suppl.* 2, 323-329, 1931.

THE MECHANICS OF GEOLOGY 159

probably never more than 1 km. deep and possibly much less.

Isostasy gives a partial but unsatisfactory explanation. Loading by sediments would produce outflow in the lower layer and therefore a lowering of the original surface. If it sank as fast as the thickness of sediments increased the depth of the sea would remain constant. But this would not be true, as A. Morley Davies and E. M. Anderson have pointed out; as the density of the sediments is less than that of the lower layer the lowering of the original floor would be less than the thickness of the sediments; 10 km. of sediments would fill up a sea 3 km. deep, and such an original depth seems to be inadmissible.

The effect of thermal blanketing, on the other hand, helps considerably. It would increase the temperatures at all depths, and in particular those in the intermediate layer. Thus in time the intermediate layer would become weaker, and the load on it would squeeze it out horizontally like the jam in a carelessly handled sandwich cake. The isostatic movement is therefore twofold. The first stage is outflow in the lower layer; but when this is complete the upper layers are still under stress-differences comparable with the load, and any weakening in them will produce further outflow. Compensation at this stage will be at the expense of the intermediate layer instead of the lower one, and the difference of density is much reduced. It appears that if this happens 10 km. of sediments can be laid down in water whose original depth was only 1·6 km., which is fairly reasonable. A complete theory, however, would have to explain also how the original depth of 1·6 km. came about; the original difficulty is reduced but not removed.

This outflow in the intermediate layer probably plays an important part in the later history of mountains. The internal movement in the mountains of Scotland and Norway is as strong as in the Alps, but the mountains themselves are much lower. The intensity of the original movement suggests that the heights of these old ranges were once as great as those of the Alps. Before geologists came to recognize isostasy as an important influence there was no difficulty about explaining how mountains come to be lowered in the course of time; they were merely supposed to have been worn down by the action of rain and rivers. But it appears now that such wearing at the surface reduces the load, and new matter comes in below, thus raising the surface afresh. To lower a mountain system by 3 km. would require the removal of about 14 km. of rock. This would remove the whole of the sediments, on any reasonable supposition about their thickness, and also most of the granitic layer. An old mountain range should therefore contain no sediments, and the granitic and even possibly the intermediate layer should be exposed everywhere within it. Actually of course these old ranges are mainly composed of sediments, much of these having been laid down not long before the mountains were uplifted. Denudation and ordinary isostasy together therefore are hard to reconcile with the presence of sediments in the old mountains. But if we allow for the possibility of outflow in the intermediate layer we arrive at an explanation. The extra load due to a new mountain system begins by producing outflow in the lower layer until compensation is established, but stress-differences tending to produce outflow persist in the granitic and intermediate layers. These can squeeze out the intermediate layer, while compensating inflow takes place in the lower layer. The chief cause of the lowering of mountains is then probably not surface denudation, but the squeezing out of the intermediate layer. This does not imply any lack of strength, because it is only when the load approaches the strength that the outflow can proceed. It appears that when the strength is exceeded by the stress-differences the geological time-scale requires a viscosity of between 10^{22} and 10^{23} c.g.s. units in the intermediate layer. This is an enormous stiffness by ordinary standards. But if the lower layer has such a viscosity compensation can be established in some tens of thousands of years. The difference in the times needed to squeeze out the intermediate layer and the lower layer under the same surface load is not chiefly a matter of difference in viscosity but of depth. In a deep layer the deformation can occur at any depth and in fact will occur down to a depth a few times the horizontal extent of the load; but in a shallow one the motion is more analogous to that of water in a capillary tube. Actually the lower layer, at least in its upper parts, probably has a higher viscosity than the intermediate one under equal stress, because it is made of less fusible materials.

In general the hypothesis that the formation of mountains is due to a contraction of the interior leads to a very satisfactory co-ordination of the facts. The explanation of the contraction itself, however, is more difficult. The only cause that has survived criticism is the general cooling of the earth, but this " thermal contraction theory " has been repeatedly attacked. On the whole the force of the objections has diminished with increasing knowledge, but the theory in its present form suffers from an indefiniteness that is likely to be very difficult to remove. On the theory of the earth's thermal history given in Chapter V the cooling at any depth consists of two parts. The first is confined to a depth of about 600 km., and leads to a total contraction whose amount since the beginning increases like the square root of the time. The second, due to the variation of the original melting-point with depth, leads to a cooling that is probably of the same order of magnitude at all depths. Near the surface it is less than that due to the first part, but it extends all the way to the centre instead of being confined to the outer tenth of the radius. The total contractions up to the present time due to the two causes seem to be comparable in amount; but the second part increases in proportion to the time instead of to the square root of the time. Now the intervals between consecutive stages of mountain formation should correspond to the accumulation of enough contraction to give fracture. If the earth is contracting at a uniform rate these intervals should all be equal, but if the rate decreases as time goes on the intervals will become longer. The first part of the contraction, if it is the more important, should imply increasing intervals between the great intervals of mountain formation; the second part would leave them nearly equal. From Holmes's analysis it seems that the second alternative agrees better with actual determinations of geological time. Holmes actually infers that the intervals have decreased, but I am inclined to think that this is merely because the more recent movements have been more thoroughly investigated. So far as I can see no existing theory would account for a decrease in the intervals, Holmes's own being no exception. On the face of it the evidence seems consistent with the idea that the contraction during geological time has mainly been due to the cooling at great depths, whereas the chief contribution from the changes of temperature down to 600 km. was confined to the early Pre-Cambrian. The difficulty in testing the theory in this form is that the amount of contraction depends on the coefficient of thermal contraction at great depths, and to a less extent on the conductivity. At present we have no knowledge of how these properties behave at high temperature and pressure, and the theory cannot be completed till we have much more knowledge of the properties of matter at high pressures than is now ready. By simply assuming that they are the same as in the laboratory we get an estimate of the contraction in the last 1500 million years that is of the right order of magnitude to account for about six major epochs of mountain formation, so that at least the theory is not obviously wrong and remains the best available until some alternative can be shown to fit the facts as well or better.

Until recently the theory of thermal contraction has always been worked out on the supposition of symmetry. There are strong reasons for believing that this is too simple, and that the cooling has varied considerably from place to place. The apparent absence of the granitic layer from the Pacific floor is perhaps the most striking instance; for in the continents this layer provides nearly half the heat outflow at the surface. But even within the continents the temperature gradient in mines varies from place to place. We do not know enough about the variation of radioactivity to produce a theory capable of detailed comparison with observation, but we can obtain suggestive results in general terms. We should expect, for instance, that at depths down to 600 km. or so the cooling since the earth solidified has been considerably more below the oceans than below the continents. An immediate consequence is that the rocks should be stronger below the Pacific than below the continents. This is verified by the distribution of the mountains around the Pacific. We have a practically continuous chain or series of chains along the western coasts of North and South America. This would be expected if the ocean floor is the stronger, because when contraction led to yield the strength would be reached first in the weaker place, and the continental rocks would be crumpled along the ocean margin.

Any difference of cooling would lead also to a difference of contraction. A body cooling uniformly and left to itself contracts in the same ratio in all directions. Where cooling is not uniform the body is pulled out of shape by an amount that has to be determined by the theory of elasticity. When a book is held in front of a fire the exposed side loses water and contracts, and the result of the difference of contraction on the two faces is that the edges curl towards the fire. Thermal contraction can produce similar effects. But as the differences are in any case confined to a depth of about a tenth of the radius any distortion is resisted by the rigidity of the interior. The result is that the inner nine-tenths of the radius is hardly distorted; the whole of the volume contraction in the outer tenth is pushed into the vertical direction, so that there is little change of dimensions in the two horizontal directions, while the contraction in the vertical is three times what it would be if unconstrained. The numerical results are affected by our almost total ignorance of the radioactivity of oceanic rocks, but it appears that with reasonable variation of the cooling, assuming that the theory of elasticity continued to hold, some places could have sunk 10 or 20 km. more than others. The stress-differences arising would, however, exceed the strength several times, and readjustment would take place at all depths in the cooled layer. This offers a possible explanation of deep-focus earthquakes. There would be little actual change of level in the readjustment; the tendency would be to restore isostasy, which is already nearly satisfied, but not quite, by the elastic theory.

These results may lead to an explanation of one of the more difficult problems of geology, that of the foundering of continents and land bridges. It appears that most of the sediments of the north and west of the British Isles and of Scandinavia were not formed from Europe, but from a continent that used to occupy the North Atlantic and sank below sea-level about the end of the Cretaceous period. The present depth between Britain and Greenland averages about 2 km. Denudation might explain lowering nearly to sea-level, but not below sea-level. The quantity of water on the surface has probably increased considerably during geological time, owing to emission from volcanoes, and this must be taken into account in any full solution, but it seems unlikely to explain the whole of the facts. The distribution of animals and plants in the past, both on land and in the ocean, seems to imply former land connections across the oceans, which both connected the faunas and floras of the continents at their ends and separated those of the seas on their opposite sides. These " land bridges " have somehow sunk below the sea and can now be traced, if at all, only by detailed measurements of depth. Such changes of level could be explained by differences of cooling, but a proper test of the theory would require measurements of the radioactivity and the temperature gradient below the present oceans, and the data are at present difficult enough to obtain within the continents. There seems to be also some possibility that differential cooling may provide the initial difference of level needed to start a geosyncline.

We cannot hope to explain all the earth's surface features in detail; even if we had all the essential data, the observed facts are so complicated that it would probably take several lifetimes to work out the consequences. We may, however, reasonably hope for a theory that will represent the facts in general terms; then we can leave the details to the geologist. It appears to me that the thermal contraction theory gives the right amount of latitude.

The origin of the distinction between continents and oceans remains very imperfectly explained; how did all the granite come to be concentrated in less than a third of the surface? If we include foundered continents in the original land surface the ratio can hardly be increased to more than a half. It seems clear that this separation could not have arisen while the viscosity was small, for then the fluid would necessarily settle down till the surfaces of equal density and pressure coincided, and there would be no room for variation in the thickness of a surface layer, much less for the complete concentration of such a layer in certain regions. It is equally clear that the separation could not have originated after a finite strength had been acquired, for that would mean that an originally uniform surface layer had to be broken up and carried to limited regions against both gravity and its own strength, and no forces are known that would be adequate or of the right kind to produce such an effect. With the right degree of viscosity, however, such separation could happen during solidification. In a liquid cooling from the top currents are set up, rising in some places and descending in others. In suitable conditions such currents can be quite steady. They usually play a greater part in the upward transfer of heat than direct conduction does. If the temperature gradient is just sufficient to excite them the vertical velocity is in the same sense at all depths. At the surface the motion is from the places where the liquid is rising to those where it is sinking. If any material separates at the surface at this stage it will be carried as a scum to the latter places and will therefore collect in separate patches. G. F. S. Hills * has suggested that the formation of continents may be accounted for on these lines. An earlier and rather similar idea was given by J. Geszti,† but he supposed the currents, if I

* *Geol. Mag.*, **71**, 275-280, 1934.
† *Gerlands Beiträge*, **27**, 1-25, 1930 ; **31**, 1-39, 1931.

understand him correctly, to be in the still unsolidified granitic layer. This seems to me to be less satisfactory, because in all the problems of this type that have been investigated the distance between the rising and sinking currents is about 1·5 times the depth of the layer. If the active layer was only the granitic one, or even the whole of the upper ones together, we might be able to account for islands about 100 km. across scattered over the whole earth, but it seems very unlikely that we could account for Asia and the Pacific Ocean. If, however, the convection currents existed throughout the rocky shell this difficulty disappears. It does, however, raise a further one: it seems to require the upper layers to have solidified before the less fusible lower one. This is less serious than might appear. The type of convection required is the simplest possible, and could occur only when the material is becoming so stiff that convection is on the verge of being stopped altogether. In these conditions the heat carried into the upper layer, even with allowance for convection, might be less than that radiated from the free-surface, and then the upper layer would solidify while the lower one, though stiff, was not definitely solid. Accordingly I think that Hills's theory is very promising, though it still requires a full investigation.

Holmes, in some of his recent work, supposes that these convection currents in the lower layer still exist and are maintained by radioactivity.* For some reason that I have not succeeded in understanding

* *J. Wash. Acad. Sci.*, 23, 169-195, 1933.

he refuses to admit that the tendency of radioactive matter to rise to the top, which he himself showed in the Finland granites, has a general application. He regards the deep-seated stress differences indicated by the distribution of gravity as incidental to the distortion involved in the convection currents. This would be qualitatively possible. He appeals to the stresses produced by the horizontal currents at the base of the upper layers to produce yield in the latter, thus accounting for mountain formation and continental drift. The interest of his theory, in my opinion, is that it shows what kind of assumptions are necessary if we insist on avoiding the view that radioactivity is sufficiently concentrated to the top to permit the earth to cool. For this reason a full quantitative examination of Holmes's theory would be worth while; but it seems to be beyond the range of our present knowledge of hydrodynamics, and may require also a much more complete theory of viscosity at high temperatures and pressures than now exists. There are, however, a few obvious difficulties. When a kettle of water is heated the motion has not the regular pattern that exists when instability has just arisen; if the heat supply is too great further instabilities arise and the result is a quite irregular system of currents. The velocity at any given point of the surface is continually changing with time, and at any moment the velocities at points separated by a small fraction of the depth may be opposite in direction. In such conditions there would be no uniformity of the stress over continental areas and for geological periods such as Holmes requires to explain geological movements. For his theory to work the supply of heat in the lower layer must be enough to produce the first instability but not the second; and in a deep layer this condition seems likely to restrict its amount to a very narrow range. Nor can I see anything in the theory to explain the observed intermittence of mountain formation. The stresses it implies act all the time, unlike those involved in the contraction theory, which are relieved at every yield; thus they would apparently either produce mountains all the time or not at all.

Holmes was, of course, the originator of the fundamental principles of the theory of the earth's thermal history that I have developed, but he abandoned this theory in 1923 on grounds that I have always considered inadequate; and every relevant piece of evidence that has arisen later seems to confirm his earlier theory and to make any alternative more difficult to maintain.

There is an obvious resemblance between Hills's theory of the origin of continents and Holmes's theory of the earth's present state, but even if either is acceptable by itself I doubt very much whether it would be possible to maintain both. The initiation of convection currents is due to instability; that is, a fluid layer heated below could remain motionless for ever, provided the temperature was a function of height alone, but with the slightest disturbance the inequalities will increase until the convective circulation is fully developed. A slight reduction of density in any vertical column, for instance, produces a rising current which draws up heated fluid there from the bottom, thus reducing the density further. There is no reason why the rising currents should be in any one place rather than any other, nor need they keep to the same places after they are formed. On Hills's theory the situations of the continents are a matter of chance; the strength of the crust keeps them where they were formed, but their original location is a matter of small disturbances beyond the range of investigation. But when continents are once formed their radioactivity implies a permanent disturbance of the thermal symmetry, and if the lower layer is weak this will give currents outwards. Similarly the currents will be outward from a heated geosyncline. Thus they are in the wrong direction to account for mountain formation, which requires the crust near a geosyncline to be pushed inwards. I think therefore, that Holmes's theory is inconsistent either with that of Hills or with geology, and as between Holmes's theory and Hills's I prefer the latter because there seems to be more need for it.

Large areas of the continents have been above sea-level since the early Pre-Cambrian, notably much of Canada, Siberia, Africa, and Australia. Other parts of the present continents have been under the sea at one time or another, and much of the present sea has been land. The original continents and oceans overlapped the present ones considerably, but not entirely. The changes may have been due largely to differences of radioactivity, and to the consequent tendency of certain regions to sink and others to rise, relatively to the mean. If this is right the original differences of level must have been much less than the present ones, and the existence of continents at an early stage is due to the amount of water on the surface having been less than it is now.

Other changes are possible that would give rise to alterations of level; change of state is one. We may note that no physical evidence supports the view of Wegener and others that the continents can drift horizontally. Such motion is opposed by the strength of the matter near the surface, which is open to no doubt whatever, and it is much easier to explain vertical than horizontal movement on a continental scale. Further, the largest force available, such as it is, is towards the equator, and would have brought the continents into a belt around the equator. Wegener's view that it would give motion to the west would require so low a viscosity that the earth would behave as fluid to tidal forces, and there would be no ocean tides.

Mountain formation is often likened to the shrivelling of the skin of an apple. The interior loses water by evaporation, and shrinks in volume, and the skin accommodates itself to the interior by a general puckering. This is such a natural analogy that the reader must be warned that it is very incomplete. It is not fundamentally a problem of fracture at all, but of what is known in engineering as elastic instability. The simplest case of this can be demonstrated by holding a postcard lengthwise between the hands and then trying to bring the hands together. So long as the force applied to the ends is not too great the card remains plane; there is some elastic compression, but not enough to be visible. But under a sufficient force the card suddenly bends into a deep arch. Take off the force and the card returns to its original plane form. The whole movement is therefore purely elastic. The fact is that for any stress the plane form is a form of equilibrium. At small stresses any small disturbance of the conditions gives only a small displacement; in other words, the plane form is stable. But at a sufficient stress the slightest disturbance gives a displacement which proceeds to grow until the form is seriously altered. This is what we mean when we say that the plane form is now unstable. The instability need not give fracture, but in some bodies the displacement may lead to fracture when it becomes great enough; if we try the same experiment on a piece of common cardboard it will probably crack. On the other hand, a thick bar squeezed on the ends will never become unstable, because it will break first.

This elastic instability is a well-known cause of danger in engineering. Thus a shaft carrying a flywheel becomes unstable and bends violently if the rotation is too fast; the next stage is that the shaft breaks and the flywheel proceeds to some distant part

THE MECHANICS OF GEOLOGY

of the building. Such shafts have, therefore, to be carefully designed so that they will not become unstable at the rate of rotation to be put on them.

The slender North American tree known as the white birch appears to provide another instance. When it becomes too tall it bends over under its own weight, unbroken but undignified. The same may be true of the bramble, which bends over till the end reaches the ground and forms a new root, to the danger of hasty walkers.

It appears, according to Southwell, that it is only bodies of exceptional form that can become elastically unstable. It is necessary that the length must be many times the thickness. In all the engineering problems the tendency to bend is resisted by the rigidity, and the thinner the specimen the less is this tendency to recovery. In the shrivelled apple there is no fracture of the skin: instability is made possible by the fact that the thickness of the skin is small compared with the circumference. In the earth's crust, however, instability is opposed both by rigidity and by gravity. It turns out that it cannot arise before fracture if the solid crust is more than a few metres thick, and therefore it is not of importance in mountain formation. Fracture would always arise first. This consideration also excludes the "tetrahedral theory", according to which the crust accommodated itself to the contraction of the interior by elevations in four regions corresponding to the Antarctic Continent and three around Asia and

North America. It is possible, however, that some shallow contortions in alluvium are due to elastic instability.*

Volcanoes and igneous intrusions are among the earth's most striking features. Rock material in these has been forced up from a considerable depth in a liquid state, and has solidified again by cooling at the surface or at a small depth. Our theory provides a natural explanation of these phenomena. The temperature at the base of the intermediate layer, in average continental conditions, is almost enough to melt basalt with its actual water content. Any disturbance of conditions tending to raise the temperature locally will therefore be liable to produce fusion. Such a disturbance is to be found in mountain formation or in sedimentation, which produce thickening of the crust and therefore blanketing; but there are probably others. Now fusion produces an increase of volume, and this may be enough to burst the upper rocks and enable the magma to pour out on the surface or at least to accommodate itself near the surface. When the magma fails to reach the surface and accumulates instead in a mass at some depth, which may be exposed later by erosion, we have what geologists call a bathylith, a laccolith or a sill, according to its form. These intrusions may be of enormous extent; a bathylith in British Columbia can be traced for over a thousand miles, and the Whin Sill underlies most of Northumberland

* Cf. *Geol. Mag.*, **69**, 321-324, 1932.

THE MECHANICS OF GEOLOGY 177

and Durham. If magma reaches the surface we have true volcanic activity. The magma may come up through pipes or through long fissures. When the activity is over these remain filled with igneous rock; a pipe gives a neck and a fissure a dyke. In some cases a nearly circular piece of the surface has fallen in and the magma has overflowed around the margin, giving what is called a ring dyke several miles across. Such ring dykes are responsible for Glencoe, Ben Nevis, and other extinct volcanoes of Scotland.

An immediate question that arises in the case of basaltic intrusions and effusions is, what produced the pressure necessary to lift them to the surface or nearly to it? Basalt is about 10 per cent. denser than granite. The pressure at the fused region would be simply that due to the weight of the layers above it, and would be balanced by that due to a smaller depth of basalt, so that a column of basalt would apparently not be raised to reach the surface. But two other factors help the uplift. Rocks expand in melting by about 10 per cent. of their volume, so that the densities of solid granite and liquid basalt are not very different. A column of liquid basalt originating near the base of a basaltic layer would be balanced against the unfused basalt and rocks intermediate between granite and basalt at higher levels, and could easily be driven up to the surface. The other consideration comes from the water content. At high pressures a considerable percentage of water is held in the basalt, but in a magma rising to the surface this is liberated and the result is a froth. Water and chemically active gases are freely liberated in the crater of Kilauea. This effect reduces the density further. It helps to explain how not only basalt, but the still denser rocks peridotite and dunite have been raised to the surface in places, though much less frequently.

We do not know yet how closely the igneous rocks visible at the surface resemble the undisturbed rocks that they were made from. Apart from the loss of volatile constituents into the air there was probably a change of composition during crystallization owing to the settling of the denser constituents. Basalt is the commonest igneous rock on the surface, but it is still far from certain that its constituents were in the form of basalt before they were fused. It is very unlikely that peridotites and dunites found at the surface represent any deep-seated layer; they appear to be the result of the settling of the denser constituents from a basalt magma in process of crystallizing.

Volcanoes seem to have some relation to ordinary mountains. The active volcanoes of the present time on the land are all in association with mountain ranges of the folded type, and such ranges as the Rocky Mountains, the Alps and the Andes all contain either active volcanoes or extinct ones. As we have seen that excess heating is to be expected in mountainous regions both before and after the uplift, this fact is explained. Volcanoes in oceanic regions, however, may depend for their existence on some

THE MECHANICS OF GEOLOGY

other factor. There is also some relation between volcanoes and earthquakes, but it seems to be mainly indirect. Both volcanoes and earthquakes are associated with mountain ranges, and therefore tend to occur in neighbouring places, but there is little or no relation between them in time. Great earthquakes in Italy are not conspicuously associated with eruptions of Etna and Vesuvius, nor are those of Japan and Mexico more frequent when the neighbouring volcanoes are active. This fact is surprising at first sight. The formation of a fissure or a pipe means a fracture and should generate elastic waves; their absence or smallness means apparently that the volcano does not make new fractures of the dimensions of those in large earthquakes. Small earthquakes, felt at distances up to a few miles, are frequent in volcanic regions, but seem to be explicable as due to the load on the crust produced by the weight of the volcano. Perhaps the original formation of a volcano, and almost certainly the formation of a fissure leading to a dyke, gave rise to major earthquakes, but it seems that present volcanic activity is simply making use of old channels.

INDEX

ADAMS, F. D., 13.
Adams, L. H., 44.
Age of Earth, 97, 99, 101, 105, 112.
Airy, Sir G. B., 88.
Alps, 71, 153.
Anderson, E. M., 127, 128, 151, 159.
Andrade, E. N. da C., 19.
Aston, F. W., 100, 106, 129.
Atmosphere, origin of, 129.

BACK-FOLDING, 155.
Bailey, E. B., 154.
Basalt, 39, 42, 94.
Bastings, L., 38.
Blanketing, 157.
Bouguer anomaly, 70, 77.
Bowie, W., 75.
Bridgman, 19.
Bullen, K. E., 131.
Butcher, J. G., 21.

CLAIRAUT, 82.
Collet, 154.
Coker, E. G., 13.
Compensation, 75.
Conrad, V., 44, 53.
Continents, origin of, 167.
Continuous distortion, 9, 12, 153.
Contraction, 162.
Convection currents, 112, 168.
Core, 38, 79.
Crustal shortening, 145.
Crystals, 11, 16, 22.

DARWIN, Sir G. H., 21, 85, 141.
Davies, A. Morley, 159.

Deep-focus earthquakes, 48, 94, 167.
Density, 79.
De Sitter, 79.
Distortion, 6.
Dubey, V. S., 105.
Dunite, 42.

EARTH, form of, 61.
— origin of, 107.
Earthquakes, 26.
— aftershocks, 54.
— distribution, 52
— extent of movement, 59.
— felt movement, 56.
— foreshocks, 54.
Eclogite, 132.
Elastic after-working, 15, 24, 54.
Elasticoviscosity, 21.
Elles, G. L., 158.
Ellipticity, 65, 78.
Epicentre, 48.

FERMOR, 42.
Fiji, 40.
Focal depth, 48.
Focus, 48.
Folding, 153.
Fracture, 27, 51, 151.

GARNET, 158.
Geodesy, 64.
Geoid, 67.
Geological time-scale, 102.
Geopotential, 67.
Geosynclines, 157, 167.
Geszti, J., 168.

Gibson, R. E., 44.
Glasses, 22.
Goldschmidt, V. M., 130, 131.
Goranson, R. W., 128.
Grace, S. F., 135.
Granite, 39, 42, 94.
Gravity, 62.
— at sea, 69, 74, 80.
Gutenberg, 38, 46.

HAYFORD, J. F., 75.
Height, 66.
Heim, 154.
Heiskanen, W., 75, 80.
Hill-creep, 151.
Hills, G. F. S., 168.
Himalayas, 88, 92, 148.
Holmes, A., 101, 105, 163, 169.
Hopkins, W., 9.
Hunter, J. de G., 77.

IGNEOUS activity, 129, 176.
Imamura, 54.
Instability, elastic, 173.
— thermal, *see* Convection.
Intermediate layer, 44, 159, 160.
International Seismological Summary, 49.
Isostasy, 75, 88.

JEANS, Sir J. H., 108.

KILAUEA, 127, 178.
Klippes, 155.
Kovarik, A. L., 151.

LATITUDE, 63.
Lehmann, I., 131.
Level surfaces, 70.
Liquids, definition, 25.
— structure of, 16.
Love, A. E. H., 30, 36, 136.

MARSHALL, P. 40.
Maxwell, 21.

Meinesz, Vening, 69, 74, 76, 93.
Metamorphism, 158.
Meteorites, 130.
Milligal, 71.
Mises, R., 9, 21.
Mohorovičić, A., 40.
Moon, 79, 80.
Mountains, distribution in time, 163.
— lowering, 160.
— origin, 144.
— Pacific, 164.

NAPPES, 154.
Nölke, F., 142.

OCEAN deeps, 76.
Oldham, R. D., 30, 35, 37.
Olivine, 42, 79.

PACIFIC mountains, 164.
— rocks, 40, 47.
Periodicity, 53.
Phillips, D. W., 15, 54.
Poisson, 30.
Pratt, 88.
Precession, 81.
Pressure, high, 13.
Pulses, 27.

QUINNEY, H., 10.

RADIOACTIVITY, 99.
Rayleigh, Lord, 29.
Read, H. H., 104.
Reuss, 21.
Rigidity, 8.
Rosenhead, L., 137.
Russell, H. N., 106.
Rutherford, Lord, 106.

SCHARDT, 153.
Scotland, 104, 154.
Scrase, F. J., 50.
Sea-level, 62.

INDEX

Sedimentary layer, 98.
Seismographs, 31.
Shear, 8.
Sodium in sea, 97.
Solid, definition, 25.
Southwell, R. V., 175.
Staub, 154.
Stechschulte, 50.
Stokes, Sir G. G., 68.
Stoneley, R., 36, 47, 49.
Strain, 3.
Strength, 9, 87, 94.
Stress, 3.
Stress-difference, 9, 84.
Suess, E., 39.
Sulphides, 130.
Suyehiro, K., 58.

TACHYLYTE, 44.
Taylor, G. I., 10, 21, 139.
Temperature, 82, 88.
— and strength, 20.
Tetrahedral theory, 175.

Thermal contraction, 162.
Tibet, 150.
Tidal friction, 99, 137.
Tides, 134.
Tilley, C. E., 158.
Turner, H. H., 29, 48.

UNWIN, 94.
Upper layers, 39, 132.

VARIATION of latitude, 135.
Viscosity, 8, 18, 161.
Volcanoes, 176.

WADATI, K., 49.
Wager, L. R., 149.
Water, origin of, 129.
Waves, 27.
Wegener, A., 90, 173.
Whin Sill, 105, 176.
Wildt, R., 130.

X-RAYS and crystals, 17.

PRINTED IN GREAT BRITAIN AT THE UNIVERSITY PRESS, ABERDEEN

METHUEN'S
GENERAL LITERATURE

A SELECTION OF

MESSRS. METHUEN'S PUBLICATIONS

This Catalogue contains only a selection of the more important books published by Messrs. Methuen. A complete catalogue of their publications may be obtained on application.

ABRAHAM (G. D.)
  MODERN MOUNTAINEERING
    Illustrated. 7s. 6d. net.
ARMSTRONG (Anthony) ('A. A.' of Punch)
  WARRIORS AT EASE
  WARRIORS STILL AT EASE
  SELECTED WARRIORS
  PERCIVAL AND I
  PERCIVAL AT PLAY
  APPLE AND PERCIVAL
  ME AND FRANCES
  HOW TO DO IT
  BRITISHER ON BROADWAY
  WHILE YOU WAIT
    Each 3s. 6d. net.
  LIVESTOCK IN BARRACKS
    Illustrated by E. H. SHEPARD.
      3s. 6d. net.
  EASY WARRIORS
    Illustrated by G. L. STAMPA.
      5s. net.
  YESTERDAILIES. Illustrated.
      3s. 6d. net.
BALFOUR (Sir Graham)
  THE LIFE OF ROBERT LOUIS STEVENSON 10s. 6d. net.
    Also, 3s. 6d. net.
BARKER (Ernest)
  NATIONAL CHARACTER
      10s. 6d. net.
  GREEK POLITICAL THEORY
      14s. net.
  CHURCH, STATE AND STUDY
      10s. 6d. net.
BELLOC (Hilaire)
  PARIS   8s. 6d. net.
  THE PYRENEES 8s. 6d. net.

BELLOC (Hilaire)—continued
  MARIE ANTOINETTE 18s. net.
  A HISTORY OF ENGLAND
    In 7 Vols. Vols. I, II, III and IV
      Each 15s. net.
BINNS (L. Elliott), D.D.
  THE DECLINE AND FALL OF THE MEDIEVAL PAPACY. 16s. net.
BIRMINGHAM (George A.)
  A WAYFARER IN HUNGARY
    Illustrated. 8s. 6d. net.
  SPILLIKINS : ESSAYS 3s. 6d. net.
  SHIPS AND SEALING-WAX : ESSAYS
      3s. 6d. net.
  CAN I BE A CHRISTIAN ? 1s. net.
CASTLEROSSE (Viscount)
  VALENTINE'S DAYS
    Illustrated. 12s. 6d. net.
CHALMERS (Patrick R.)
  KENNETH GRAHAME : LIFE, LETTERS AND UNPUBLISHED WORK
    Illustrated. 10s. 6d. net.
CHARLTON (Moyra)
  PATCH : THE STORY OF A MONGREL
    Illustrated by G. D. ARMOUR.
      2s. 6d. net.
  THE MIDNIGHT STEEPLECHASE
    Illustrated by GILBERT HOLIDAY.
      5s. net.
CHESTERTON (G. K.)
  COLLECTED POEMS 7s. 6d. net.
  ALL I SURVEY   6s. net.
  THE BALLAD OF THE WHITE HORSE
      3s. 6d. net.
  Also illustrated by ROBERT AUSTIN. 12s. 6d. net.

Messrs. Methuen's Publications

CHESTERTON (G. K.)—*continued*
  ALL IS GRIST
  CHARLES DICKENS
  COME TO THINK OF IT ...
  GENERALLY SPEAKING
  ALL THINGS CONSIDERED
  TREMENDOUS TRIFLES
  FANCIES VERSUS FADS
  ALARMS AND DISCURSIONS
  A MISCELLANY OF MEN
  THE USES OF DIVERSITY
  THE OUTLINE OF SANITY
  THE FLYING INN
      Each 3s. 6d. net.
  WINE, WATER AND SONG 1s. 6d. net.

CURLE (J. H.)
  THE SHADOW-SHOW   6s. net.
    Also, 3s. 6d. net.
  THIS WORLD OF OURS   6s. net.
  TO-DAY AND TO-MORROW   6s. net.
  THIS WORLD FIRST   6s. net.

DUGDALE (E. T. S.)
  GERMAN DIPLOMATIC DOCUMENTS, 1871–1914
  In 4 vols. Vol. I, 1871–90. Vol. II, 1891–8. Vol. III, 1898–1910. Vol. IV, 1911–14.
      Each £1 1s. net.

EDWARDES (Tickner)
  THE LORE OF THE HONEY-BEE
  Illustrated. 7s. 6d. and 3s. 6d. net.
  BEE-KEEPING FOR ALL
    Illustrated. 3s. 6d. net.
  THE BEE-MASTER OF WARRILOW
    Illustrated. 7s. 6d. net.
  BEE-KEEPING DO'S AND DON'TS
    2s. 6d. net.
  LIFT-LUCK ON SOUTHERN ROADS
    5s. net.

EINSTEIN (Albert)
  RELATIVITY : THE SPECIAL AND GENERAL THEORY   5s. net.
  SIDELIGHTS ON RELATIVITY
    3s. 6d. net.
  THE MEANING OF RELATIVITY
    5s. net.
  THE BROWNIAN MOVEMENT
    5s. net.

EISLER (Robert)
  THE MESSIAH JESUS AND JOHN THE BAPTIST
    Illustrated. £2 2s. net.

EWING (A. C.)
  IDEALISM   21s. net.

FIELD (G. C.)
  MORAL THEORY   6s. net.
  PLATO AND HIS CONTEMPORARIES
    12s. 6d. net.
  PREJUDICE AND IMPARTIALITY
    2s. 6d. net.

FINER (H.)
  THE THEORY AND PRACTICE OF MODERN GOVERNMENT   2 vols.
    £2 2s. net.
  ENGLISH LOCAL GOVERNMENT
    £1 1s. net.

FYLEMAN (Rose)
  HAPPY FAMILIES
  FAIRIES AND CHIMNEYS
  THE FAIRY GREEN
  THE FAIRY FLUTE   *Each* 2s. net.
  THE RAINBOW CAT
  EIGHT LITTLE PLAYS FOR CHILDREN
  FORTY GOOD-NIGHT TALES
  FORTY GOOD-MORNING TALES
  SEVEN LITTLE PLAYS FOR CHILDREN
  TWENTY TEA-TIME TALES
    Each 3s. 6d. net.
  THE BLUE RHYME BOOK
    Illustrated. 3s. 6d. net.
  THE EASTER HARE
    Illustrated. 3s. 6d. net.
  FIFTY-ONE NEW NURSERY RHYMES
    Illustrated by DOROTHY BURROUGHES.   6s. net.
  THE STRANGE ADVENTURES OF CAPTAIN MARWHOPPLE
    Illustrated. 3s. 6d. net.

GIBBON (Edward)
  THE DECLINE AND FALL OF THE ROMAN EMPIRE
  With Notes, Appendixes and Maps, by J. B. BURY. Illustrated. 7 vols. 15s. net each volume. Also, unillustrated, 7s. 6d. net each volume.

GLOVER (T. R.)
  VIRGIL
  THE CONFLICT OF RELIGIONS IN THE EARLY ROMAN EMPIRE
  POETS AND PURITANS
    Each 10s. 6d. net.
  FROM PERICLES TO PHILIP
    12s. 6d. net.

Messrs. Methuen's Publications

GRAHAME (Kenneth)
  THE WIND IN THE WILLOWS
    7s. 6d. net and 5s. net.
  Also illustrated by ERNEST H.
    SHEPARD. *Cloth*, 7s. 6d. net.
    Green Leather, 12s. 6d. net.
  Pocket Edition, unillustrated.
    Cloth, 3s. 6d. net.
    Green Morocco, 7s. 6d. net.
  THE KENNETH GRAHAME BOOK
    ('The Wind in the Willows',
    'Dream Days' and 'The Golden
    Age' in one volume).
                       7s. 6d. net.
  See also Milne (A. A.)

HALL (H. R.)
  THE ANCIENT HISTORY OF THE
    NEAR EAST   £1 1s. net.
  THE CIVILIZATION OF GREECE IN
    THE BRONZE AGE £1 10s. net.

HEATON (Rose Henniker)
  THE PERFECT HOSTESS
    Decorated by A. E. TAYLOR.
  7s. 6d. net. Gift Edition, £1 1s. net.
  THE PERFECT SCHOOLGIRL
                       3s. 6d. net.

HEIDEN (Konrad)
  A HISTORY OF NATIONAL SOCIALISM
                       15s. net.

HERBERT (A. P.)
  HELEN   2s. 6d. net.
  TANTIVY TOWERS and DERBY DAY
    in one volume. Illustrated by
    Lady VIOLET BARING. 5s. net.
  Each, separately, unillustrated
                    2s. 6d. net.
  HONEYBUBBLE & CO. 3s. 6d. net.
  MISLEADING CASES IN THE COMMON
    LAW   5s. net.
  MORE MISLEADING CASES 5s. net.
  STILL MORE MISLEADING CASES
                       5s. net.
  THE WHEREFORE AND THE WHY
  'TINKER, TAILOR ...'
    Each, illustrated by GEORGE
      MORROW.   2s. 6d. net.
  THE SECRET BATTLE   3s. 6d. net.
  THE HOUSE BY THE RIVER
                       3s. 6d. net.
  'NO BOATS ON THE RIVER'
    Illustrated.   5s. net

HOLDSWORTH (Sir W. S.)
  A HISTORY OF ENGLISH LAW
  Nine Volumes.   £1 5s. net each.
  Index Volume by EDWARD POTTON.
                    £1 1s. net.

HUDSON (W. H.)
  A SHEPHERD'S LIFE
    Illustrated. 10s. 6d. net.
  Also unillustrated. 3s. 6d. net.

HUTTON (Edward)
  CITIES OF SICILY
    Illustrated. 10s. 6d. net.
  MILAN AND LOMBARDY.
  THE CITIES OF ROMAGNA AND THE
    MARCHES
  SIENA AND SOUTHERN TUSCANY
  NAPLES AND SOUTHERN ITALY
    Illustrated. *Each* 8s. 6d. net.
  A WAYFARER IN UNKNOWN TUSCANY
  THE CITIES OF SPAIN
  THE CITIES OF UMBRIA
  COUNTRY WALKS ABOUT FLORENCE
  ROME
  FLORENCE AND NORTHERN TUSCANY
  VENICE AND VENETIA
    Illustrated. *Each* 7s. 6d. net.

INGE (W. R.), D.D., Dean of St. Paul's
  CHRISTIAN MYSTICISM. With a New
    Preface.   7s. 6d. net.

JOHNS (Rowland)
  DOGS YOU'D LIKE TO MEET
  LET DOGS DELIGHT
  ALL SORTS OF DOGS
  LET'S TALK OF DOGS
  PUPPIES
  LUCKY DOGS
    Each, Illustrated, 3s. 6d. net.
  SO YOU LIKE DOGS !
    Illustrated.   2s. 6d. net.
  THE ROWLAND JOHNS DOG BOOK.
    Illustrated.   5s. net.

'OUR FRIEND THE DOG' SERIES
Edited by ROWLAND JOHNS.
  THE CAIRN
  THE COCKER SPANIEL
  THE FOX-TERRIER
  THE PEKINGESE
  THE AIREDALE
  THE ALSATIAN
  THE SCOTTISH TERRIER
  THE CHOW-CHOW
  THE IRISH SETTER
  THE DALMATIAN
  THE LABRADOR
  THE SEALYHAM
  THE DACHSHUND
  THE BULLDOG
  THE BULL-TERRIER
  THE GREAT DANE
  THE POMERANIAN
  THE COLLIE
  THE ENGLISH SPRINGER
                Each 2s. 6d. net.

Messrs. Methuen's Publications

KIPLING (Rudyard)
  BARRACK-ROOM BALLADS
  THE SEVEN SEAS
  THE FIVE NATIONS
  DEPARTMENTAL DITTIES
  THE YEARS BETWEEN
  Four Editions of these famous volumes of poems are now published, viz. :—
              Buckram, 7s. 6d. net.
  Cloth, 6s. net. Leather, 7s. 6d. net.
  Service Edition. Two volumes each book.   3s. net each vol.
  A KIPLING ANTHOLOGY—VERSE
            Leather, 7s. 6d. net.
    Cloth, 6s. net and 3s. 6d. net.
  TWENTY POEMS FROM RUDYARD KIPLING   1s. net.
  A CHOICE OF SONGS   2s. net.
  SELECTED POEMS   1s. net.

LAMB (Charles and Mary)
  THE COMPLETE WORKS
    Edited by E. V LUCAS. Six volumes.   6s. net each.
  SELECTED LETTERS
    Edited by G. T. CLAPTON.
                3s. 6d. net.
  THE CHARLES LAMB DAY-BOOK
    Compiled by E. V. LUCAS. 6s. net.
  THE LETTERS OF CHARLES LAMB
    Edited by E. V. LUCAS. Two volumes.   6s. net each.
  THE BEST OF LAMB
    Edited by E. V. LUCAS. 2s. 6d. net.

LANKESTER (Sir Ray)
  SCIENCE FROM AN EASY CHAIR First Series
  SCIENCE FROM AN EASY CHAIR Second Series
  GREAT AND SMALL THINGS
      Each, Illustrated, 7s. 6d. net.
  SECRETS OF EARTH AND SEA
      Illustrated. 8s. 6d. net.

LENNHOFF (Eugen)
  THE FREEMASONS   21s. net.

LINDRUM (Walter)
  BILLIARDS. Illustrated. 2s. 6d. net.

LODGE (Sir Oliver)
  MAN AND THE UNIVERSE
      7s. 6d. net and 3s. 6d. net.
  THE SURVIVAL OF MAN 7s. 6d. net.
  RAYMOND   10s. 6d. net.
  RAYMOND REVISED   6s. net.
  MODERN PROBLEMS   3s. 6d. net.
  REASON AND BELIEF   3s. 6d. net.
  THE SUBSTANCE OF FAITH 2s. net.
  RELATIVITY   1s. net.
  CONVICTION OF SURVIVAL 2s. net.

LUCAS (E. V.), C.H.
  READING, WRITING AND REMEMBERING   18s. net.
  THE COLVINS AND THEIR FRIENDS
                  £1 1s. net.
  THE LIFE OF CHARLES LAMB
    2 Vols.   £1 1s. net.
  AT THE SHRINE OF ST. CHARLES
                    5s. net.
  POST-BAG DIVERSIONS 7s. 6d. net.
  VERMEER THE MAGICAL   5s. net.
  A WANDERER IN ROME
  A WANDERER IN HOLLAND
  A WANDERER IN LONDON
  LONDON REVISITED (Revised)
  A WANDERER IN PARIS
  A WANDERER IN FLORENCE
  A WANDERER IN VENICE
           Each 10s. 6d. net.
  A WANDERER AMONG PICTURES
                  8s. 6d. net.
  E. V. LUCAS'S LONDON £1 net.
  THE OPEN ROAD   6s. net.
  Also, illustrated by CLAUDE A. SHEPPERSON, A.R.W.S.
               10s. 6d. net.
  Also, India Paper.
           Leather, 7s. 6d. net.
  THE JOY OF LIFE   6s. net.
    Leather Edition, 7s. 6d. net.
  Also, India Paper.
           Leather, 7s. 6d. net.
  THE GENTLEST ART
  THE SECOND POST
  FIRESIDE AND SUNSHINE
  CHARACTER AND COMEDY
  GOOD COMPANY
  ONE DAY AND ANOTHER
  OLD LAMPS FOR NEW
  LOITERER'S HARVEST
  LUCK OF THE YEAR
  EVENTS AND EMBROIDERIES
  A FRONDED ISLE
  A ROVER I WOULD BE
  GIVING AND RECEIVING
  HER INFINITE VARIETY
  ENCOUNTERS AND DIVERSIONS
  TURNING THINGS OVER
  TRAVELLER'S LUCK
  AT THE SIGN OF THE DOVE
  VISIBILITY GOOD Each 3s. 6d. net.
  LEMON VERBENA
  SAUNTERER'S REWARDS
                Each 6s. net.
  FRENCH LEAVES
  ENGLISH LEAVES
  THE BARBER'S CLOCK Each 5s. net.
  'THE MORE I SEE OF MEN . . .'

Messrs. Methuen's Publications

LUCAS (E. V.)—*continued*
  OUT OF A CLEAR SKY
  IF DOGS COULD WRITE
  '. . . AND SUCH SMALL DEER'
                  Each 3s. 6d. *net.*
  See also Lamb (Charles).

LYND (Robert)
  THE COCKLESHELL   5s. *net.*
  RAIN, RAIN, GO TO SPAIN
  IT'S A FINE WORLD
  THE GREEN MAN
  THE PLEASURES OF IGNORANCE
  THE GOLDFISH
  THE LITTLE ANGEL
  THE BLUE LION
  THE PEAL OF BELLS
  THE ORANGE TREE
  THE MONEY-BOX   *Each* 3s. 6d. *net.*
  'YY.' An Anthology of essays by
  ROBERT LYND. Edited by EILEEN
  SQUIRE.   7s. 6d. *net.*

McDOUGALL (William)
  AN INTRODUCTION TO SOCIAL
    PSYCHOLOGY   10s. 6d. *net.*
  NATIONAL WELFARE AND NATIONAL
    DECAY   6s. *net.*
  AN OUTLINE OF PSYCHOLOGY
                   10s. 6d. *net.*
  AN OUTLINE OF ABNORMAL PSYCHO-
    LOGY   15s. *net.*
  BODY AND MIND   12s. 6d. *net.*
  CHARACTER AND THE CONDUCT OF
    LIFE   10s. 6d. *net.*
  MODERN MATERIALISM AND EMER-
    GENT EVOLUTION   3s. 6d. *net.*
  ETHICS AND SOME MODERN WORLD
    PROBLEMS   7s. 6d. *net.*
  THE ENERGIES OF MEN 8s. 6d. *net.*
  RELIGION AND THE SCIENCES OF
    LIFE   8s. 6d. *net.*

MAETERLINCK (Maurice)
  THE BLUE BIRD   6s. *net.*
  Also, illustrated by F. CAYLEY
    ROBINSON.   10s. 6d. *net.*
  OUR ETERNITY   6s. *net.*
  THE UNKNOWN GUEST   6s. *net.*
  POEMS   5s. *net.*
  THE WRACK OF THE STORM 6s. *net.*
  THE BETROTHAL   6s. *net.*
  MARY MAGDALENE   2s. *net.*

MARLOWE (Christopher)
  THE WORKS. In 6 volumes.
    General Editor, R. H. CASE.
    THE LIFE OF MARLOWE and DIDO,
    QUEEN OF CARTHAGE 8s. 6d. *net.*
    TAMBURLAINE, I AND II 10s.6d.*net.*

MARLOWE (Christopher)—*cont.*
  THE WORKS—*continued*
    THE JEW OF MALTA and THE
      MASSACRE AT PARIS 10s. 6d. *net.*
    POEMS   10s. 6d. *net.*
    DOCTOR FAUSTUS   8s. 6d. *net.*
    EDWARD II   8s. 6d. *net.*

MARTIN (William)
  UNDERSTAND CHINA   7s. 6d. *net.*

MASEFIELD (John)
  ON THE SPANISH MAIN 8s. 6d. *net.*
  A SAILOR'S GARLAND   3s. 6d. *net.*
  SEA LIFE IN NELSON'S TIME
                    7s. 6d. *net.*

METHUEN (Sir A.)
  AN ANTHOLOGY OF MODERN VERSE
  SHAKESPEARE TO HARDY: An
    Anthology of English Lyrics.
           Each, Cloth, 6s. *net.*
           Leather, 7s. 6d. *net.*

MILNE (A. A.)
  TOAD OF TOAD HALL
    A Play founded on Kenneth
    Grahame's 'The Wind in the
    Willows'.   5s. *net.*
  THOSE WERE THE DAYS: Collected
    Stories   7s. 6d. *net.*
  BY WAY OF INTRODUCTION
  NOT THAT IT MATTERS
  IF I MAY
  THE SUNNY SIDE
  THE RED HOUSE MYSTERY
  ONCE A WEEK
  THE HOLIDAY ROUND
  THE DAY'S PLAY
  MR. PIM PASSES BY
              Each 3s. 6d. *net.*
  WHEN WE WERE VERY YOUNG
  WINNIE-THE-POOH
  NOW WE ARE SIX
  THE HOUSE AT POOH CORNER
  Each illustrated by E. H. SHEPARD.
  7s. 6d. *net. Leather,* 10s. 6d. *net.*
  THE CHRISTOPHER ROBIN VERSES
    ('When We were Very Young'
    and 'Now We are Six' com-
    plete in one volume). Illustrated
    in colour and line by E. H.
    SHEPARD.   8s. 6d. *net.*
  THE CHRISTOPHER ROBIN STORY
    BOOK
    Illustrated by E. H. SHEPARD.
                      5s. *net.*
  THE CHRISTOPHER ROBIN BIRTH-
    DAY BOOK
    Illustrated by E. H. SHEPARD.
                   3s. 6d. *net.*

MILNE (A. A.) and FRASER-SIMSON (H.)
   FOURTEEN SONGS FROM 'WHEN WE WERE VERY YOUNG' 7s. 6d. net.
   TEDDY BEAR AND OTHER SONGS FROM 'WHEN WE WERE VERY YOUNG' 7s. 6d. net.
   THE KING'S BREAKFAST 3s. 6d. net.
   SONGS FROM 'NOW WE ARE SIX' 7s. 6d. net.
   MORE 'VERY YOUNG' SONGS 7s. 6d. net.
   THE HUMS OF POOH 7s. 6d. net.
   In each case the words are by A. A. MILNE, the music by H. FRASER-SIMSON, and the decorations by E. H. SHEPARD.

MITCHELL (Abe)
   DOWN TO SCRATCH 5s. net.

MORTON (H. V.)
   A LONDON YEAR
      Illustrated, 6s. net.
   THE HEART OF LONDON 3s. 6d. net.
   Also, with Scissor Cuts by L. HUMMEL. 6s. net.
   THE SPELL OF LONDON
   THE NIGHTS OF LONDON
   BLUE DAYS AT SEA Each 3s. 6d. net.
   IN SEARCH OF ENGLAND
   THE CALL OF ENGLAND
   IN SEARCH OF SCOTLAND
   IN SCOTLAND AGAIN
   IN SEARCH OF IRELAND
   IN SEARCH OF WALES
      Each, illustrated, 7s. 6d. net.

NOMA (Seiji)
   THE NINE MAGAZINES OF KODANSHA: The Autobiography of a Japanese Publisher. Illustrated. 10s. 6d. net.

OMAN (Sir Charles)
   THINGS I HAVE SEEN 8s. 6d. net.
   A HISTORY OF THE ART OF WAR IN THE MIDDLE AGES, A.D. 378–1485. 2 vols. Illustrated. £1 16s. net.
   STUDIES IN THE NAPOLEONIC WARS 8s. 6d. net.

PETRIE (Sir Flinders)
   A HISTORY OF EGYPT
   In 6 Volumes.
   Vol. I. FROM THE IST TO THE XVITH DYNASTY 12s. net.
   Vol. II. THE XVIITH AND XVIIITH DYNASTIES 9s. net.
   Vol. III. XIXTH TO XXXTH DYNASTIES 12s. net.
   Vol. IV. EGYPT UNDER THE PTOLEMAIC DYNASTY
   By EDWYN BEVAN. 15s. net.

PETRIE (Sir Flinders)—continued
   Vol. V. EGYPT UNDER ROMAN RULE
   By J. G. MILNE. 12s. net.
   Vol. VI. EGYPT IN THE MIDDLE AGES
   By S. LANE POOLE. 10s. net.

PHILLIPS (Sir Percival)
   FAR VISTAS 12s. 6d. net.

POLLOCK (William)
   THE CREAM OF CRICKET 5s. net.

QUIGLEY (H.) and GOLDIE (I.)
   HOUSING AND SLUM CLEARANCE IN LONDON 10s. 6d. net.

RAGLAN (Lord)
   JOCASTA'S CRIME 6s. net.
   THE SCIENCE OF PEACE 3s. 6d. net.

SELLAR (W. C.) and YEATMAN (R. J.)
   1066 AND ALL THAT
   AND NOW ALL THIS
   HORSE NONSENSE
   Each illustrated by JOHN REYNOLDS. 5s. net.

STEVENSON (R. L.)
   THE LETTERS Edited by Sir SIDNEY COLVIN. 4 Vols. Each 6s. net.

STOCK (Vaughan)
   THE LIFE OF CHRIST 6s. net.

SURTEES (R. S.)
   HANDLEY CROSS
   MR. SPONGE'S SPORTING TOUR
   ASK MAMMA
   MR. FACEY ROMFORD'S HOUNDS
   PLAIN OR RINGLETS?
   HILLINGDON HALL
      Each, illustrated, 7s. 6d. net.
   JORROCKS'S JAUNTS AND JOLLITIES
   HAWBUCK GRANGE
      Each, illustrated, 6s. net.

TAYLOR (A. E.)
   PLATO: THE MAN AND HIS WORK £1 1s. net.
   PLATO: TIMÆUS AND CRITIAS 6s. net.
   ELEMENTS OF METAPHYSICS 12s. 6d. net.

TILDEN (William T.)
   THE ART OF LAWN TENNIS Revised Edition.
   SINGLES AND DOUBLES
      Each, illustrated, 6s. net.
   THE COMMON SENSE OF LAWN TENNIS
   MATCH PLAY AND THE SPIN OF THE BALL. Each, illustrated, 5s. net.

Messrs. Methuen's Publications

TILESTON (Mary W.)
  DAILY STRENGTH FOR DAILY NEEDS
    3s. 6d. net.
    India Paper. Leather, 6s. net.
UNDERHILL (Evelyn)
  MYSTICISM Revised Edition.
    15s. net.
  THE LIFE OF THE SPIRIT AND THE LIFE OF TO-DAY 7s. 6d. net.
  MAN AND THE SUPERNATURAL
    3s. 6d. net.
  THE GOLDEN SEQUENCE
    Paper boards, 3s. 6d. net;
    Cloth, 5s. net.
  MIXED PASTURE : Essays and Addresses 5s. net.
  CONCERNING THE INNER LIFE
    2s. net.
  THE HOUSE OF THE SOUL 2s. net.
VIEUCHANGE (Michel)
  SMARA : THE FORBIDDEN CITY
    Illustrated. 8s. 6d. net.
WARD (A. C.)
  TWENTIETH CENTURY LITERATURE
    5s. net.
  THE NINETEEN-TWENTIES 5s. net.
  LANDMARKS IN WESTERN LITERATURE 5s. net.
  AMERICAN LITERATURE 7s. 6d. net.
  WHAT IS THIS LIFE ? 5s. net.
  THE FROLIC AND THE GENTLE : A CENTENARY STUDY OF CHARLES LAMB 6s. net.

WILDE (Oscar)
  LORD ARTHUR SAVILE'S CRIME AND THE PORTRAIT OF MR. W. H.
    6s. 6d. net.
  THE DUCHESS OF PADUA
    3s. 6d. net.
  POEMS 6s. 6d. net.
  LADY WINDERMERE'S FAN
    6s. 6d. net.
  A WOMAN OF NO IMPORTANCE
    6s. 6d. net.
  AN IDEAL HUSBAND 6s. 6d. net.
  THE IMPORTANCE OF BEING EARNEST
    6s. 6d. net.
  A HOUSE OF POMEGRANATES
    6s. 6d. net.
  INTENTIONS 6s. 6d. net.
  DE PROFUNDIS and PRISON LETTERS
    6s. 6d. net.
  ESSAYS AND LECTURES 6s. 6d. net.
  SALOMÉ, A FLORENTINE TRAGEDY, and LA SAINTE COURTISANE
    2s. 6d. net.
  SELECTED PROSE OF OSCAR WILDE
    6s. 6d. net.
  ART AND DECORATION
    6s. 6d. net.
  FOR LOVE OF THE KING
    5s. net.
  VERA, OR THE NIHILISTS
    6s. 6d. net.
WILLIAMSON (G. C.)
  THE BOOK OF FAMILLE ROSE
    Richly illustrated. £8 8s. net.

METHUEN'S COMPANIONS TO MODERN STUDIES
  SPAIN. E. ALLISON PEERS. 12s. 6d. net.
  GERMANY. J. BITHELL. 15s. net.
  ITALY. E. G. GARDNER. 12s. 6d. net.
  FRANCE. R. L. G. RITCHIE. 12s. 6d. net

METHUEN'S HISTORY OF MEDIEVAL AND MODERN EUROPE
In 8 Vols. Each 16s. net.

I.	476 to 911.	By J. H. BAXTER.
II.	911 to 1198.	By Z. N. BROOKE.
III.	1198 to 1378.	By C. W. PREVITÉ-ORTON.
IV.	1378 to 1494.	By W. T. WAUGH.
V.	1494 to 1610.	By A. J. GRANT.
VI.	1610 to 1715.	By E. R. ADAIR.
VII.	1715 to 1815.	By W. F. REDDAWAY.
VIII.	1815 to 1923.	By Sir J. A. R. MARRIOTT.

Methuen & Co. Ltd., 36 Essex Street, London, W.C.2

METAMORPHISM

By

A. HARKER, M.A., F.R.S.

With 183 *Illustrations by the author*
Demy 8vo.   17s. 6d. *net*

Although much has been written concerning metamorphosed rocks from the petrographical point of view, the actual processes of metamorphism have hitherto received much less attention. The present work is an attempt to discuss this subject on systematic lines. It is divided into two parts. The first is a study of the transformations set up in rocks when subjected merely to high temperature; the second takes account of regional metamorphism. The place of petrographical description is partly supplied by numerous illustrations, drawn from the microscope.

METHUEN

CPSIA information can be obtained
at www.ICGtesting.com
Printed in the USA
BVHW050513040222
627987BV00003B/100